My Book

This book belongs to

Cover Design by :
Gowri Vemuri

First Edition :
AUGUST, 2019

Second Edition :
FEBRUARY, 2022

Author :
Gowri Vemuri

Edited by :
Raksha Pothapragada
Ritvik Pothapragada

Questions: mathknots.help@gmail.com

NOTE : These tests are neither affiliated nor sponsored or endorsed by Reverside assesments or any organization.

Dedication

This book is dedicated to:

My Mom, who is my best critic, guide and supporter.

To what I am today, and what I am going to become tomorrow,

is all because of your blessings, unconditional affection and support.

This book is dedicated to the

strongest women of my life ,

my dearest mom

and

to all those moms in this universe.

G.V.

The Iowa Algebra Aptitude Test (IAAT) was developed to help teachers and counselors make informed decisions about initial placement of students in the advanced mathematics curriculum. IAAT test provides a four-part profile of students that identifies specific areas of strength and weakness. This test is divided into the following 4 sub-tests and total test time is 40 minutes

1) Pre-Algebraic Number Skills and Concepts

2) Data Interpretation and Analysis

3) Mathematical Relationships

4) Mathematic Symbols

Different Counties conduct these tests in different grades. As an example, the following table lists the grades in which the IAAT test is conducted in Northern Virginia.

Fairfax County Public Schools (FCPS)	Grade 6
Loudoun County Public School (LCPS)	Grade 5
Prince William County Public Schools (PWCPS)	Grade 5

This test is a time-based test, for which speed and accuracy are important factors. Students are required to get 92% and above in the IAAT test and a pass advance in the SOL's/End of year assessments at school. To help kids excel in this test, **Math-Knots** published several books.

IAAT (IOWA Algebra Aptitude) Practice tests Vol 1	6 practice tests	https://www.math-knots.com/iaatpracticetests
IAAT (IOWA Algebra Aptitude) Practice tests Vol 2	8 practice tests	https://www.math-knots.com/iaatpracticetests
IOWA Algebra Aptitude Work Book - Vol 1	1000 practice questions	https://www.math-knots.com/iaatworkbook
IOWA Algebra Aptitude Work Book - Vol 2	1000 practice questions	https://www.math-knots.com/iaatworkbook

INDEX

IAAT
PRACTICE TEST
TEST - 7
2
3
A
B
A
B
C

SECTION A

1. Rik has $\frac{3}{4}$ of the cake left from his birthday party. He wants to cut it in pieces of $\frac{1}{12}$. How many pieces can he cut ?

(A) 9 (B) 6

(C) 12 (D) 4

2. A brand new housing community is planting trees across the side walks and wants to space them equally to look nice. There are 15 plants and the length of the side walk is 154 meters . what should be the distance between any two plants ?

(A) 15 (B) 10

(C) 14 (D) 11

3. Evaluate $[(-6) - (-8)]\,[(-2) - 4]$

(A) -84 (B) -12

(C) -14 (D) 12

4. Evaluate $\frac{17.17}{1.7} = \boxed{?}$

(A) 10.1 (B) 1.11

(C) 1.001 (D) 0.101

5. Find the Least Common Multiple (LCM) of the below numbers ?

3 and 7

(A) 36

(B) 21

(C) 7

(D) 42

6. Simplify $\frac{2}{3} \div \frac{13}{7}$

(A) $\frac{7}{39}$

(B) $\frac{14}{39}$

(C) $\frac{4}{13}$

(D) $\frac{39}{7}$

7. Tej solves puzzles at an average rate of 10 per hour in 7 hours how many puzzles will he solve ?

(A) $\frac{7}{10}$

(B) 10

(C) $\frac{10}{7}$

(D) 70

8. Kiran charges $60 for every 25 Sq ft of the carpet to install. How much will it cost to install 100 sq ft of carpet in each of two rooms ?

(A) 240

(B) 120

(C) 360

(D) 480

9. A cat has 2 cats and each cat then has 4 cats each. How many legs are there all together ?

(A) 40 (B) 44

(C) 14 (D) 32

10. If the length of the rectangle is 10% more than width. If the width of the rectangle is 10 cm. What is the area ?

(A) 80 sq cm (B) 100 sq cm

(C) 120 sq cm (D) 110 sq cm

11. Lucy has 3 choices of break fast and 5 different milk shake recipes to make on the Saturday. How many different ways can she make 1 break fast and 1 milk shake ?

(A) 15 (B) 8

(C) 5 (D) 16

12. Express the below number in scientific notation.

1.46

(A) 1.46×10^{0} (B) 1.46×10^{1}

(C) 1.46×10^{-1} (D) 1.46×10^{2}

13. Solve the below inequality

$$2x - 3 > 7$$

(A) $x < 5$ (B) $x > -5$

(C) $x > 5$ (D) $x > -18$

14. Find the Greatest Common Factor for the below:

72, 27

(A) 9 (B) 8

(C) 54 (D) 27

15. Write the below percent as a fraction.

48%

(A) $\frac{12}{25}$ (B) $\frac{17}{25}$

(C) 48 (D) $24\frac{1}{2}$

SECTION B

1. Evaluate H, if $\frac{46}{H} = \frac{24}{48}$

(A) 60

(B) 23

(C) 92

(D) 46

2. Find the area of a circle "A" with a diameter of 6 cm?

(A) π sq cm

(B) 36 π sq cm

(C) 9 π sq cm

(D) 3 π sq cm

3. Evaluate K where $K + 77 = 107$

(A) 37

(B) 30

(C) 70

(D) 77

4. Solve for U and T

U + T = 5 and U - T = 1, where U and T are positive integers.

(A) -3 and -2

(B) -3 and 2

(C) 3 and 2

(D) 3 and -2

5. If $R = 8 - P$ then $\boxed{\dfrac{3(R+P)}{4} = \boxed{?}}$

(A) 2

(B) 6

(C) 3

(D) 8

6. Evaluate $T \times T + U \times U$ where $T = 5$ and $U = -2$.

(A) 21

(B) 19

(C) 32

(D) 29

7. If $K \Delta P = K \times K + P \times P$, then what is the value of $(5\Delta2) + (7\Delta1)$?

(A) 77

(B) 65

(C) 79

(D) 59

8. What percent of 57 inches is 6 inches?

(A) 950%

(B) 0.11%

(C) 9.5%

(D) 10.5%

9. If U = -2 and T = -3 then $\boxed{\frac{(4U-T)}{5} = \boxed{?}}$

(A) 1

(B) -1

(C) -2

(D) 2

10. If U = 9 and U - 2T = 10 then find the value of T ?

(A) $\frac{1}{2}$

(B) 2

(C) $-\frac{1}{2}$

(D) -2

11. Sai can solve 11 puzzles in H hours. He solves 117 puzzles in 35 hours. Using ratio and proportion which of the below statements are true ?

(A) $\frac{H}{11} = \frac{35}{117}$

(B) $\frac{H}{11} = \frac{117}{35}$

(C) $\frac{11}{H} = \frac{35}{117}$

(D) $117 \times 35 \times H = 11$

12. Express the below algebraic expression in words.

$N \cdot 9 = 19$

(A) A number to the 9th is 19

(B) 9 minus a number is 19

(C) The product of a number and 9 is 19

(D) The 9th power of a number is 19

13. Evaluate $\frac{H^3}{3}$ where $H^5 = 243.$

(A) 81

(B) 18

(C) 9

(D) 27

14. Simplify $6(-4n + 4) + 9$

(A) -14n + 6

(B) -5 - 16n

(C) -79 + 20n

(D) -24n + 33

15. Simplify $-9 \geq \frac{a}{2}$

(A) $a \geq \frac{-9}{2}$

(B) $a \leq -18$

(C) $a \leq \frac{-9}{2}$

(D) $a \geq 11$

SECTION C

1. What is the slope of the line that passes through the points P(16, 3) and Q(11, 4) ?

(A) $\frac{-1}{5}$

(B) $\frac{1}{5}$

(C) -5

(D) 5

2. If T = 5U + 7 and T = 22, what is the value of U ?

(A) $\frac{1}{3}$

(B) 3

(C) -3

(D) 6

3. Simplify 150 = 5(4x + 2)

(A) 6

(B) 16

(C) 12

(D) 7

4. The ratio of yellow rose plants to red rose plants at a local nursery are 2:1. If there are 7 red rose plants, then how many yellow rose plants are there ?

(A) 12

(B) 14

(C) 11

(D) 5

5. What is the T-intercept of $U = 5T - 110$?

(A) 22

(B) + 110

(C) $\frac{1}{5}$

(D) -110

6. Which of the below is not a function?

(A) {(1, 2), (2, 3), (3, 4), (4, 5), (5, 6)}

(B) {(1, 5), (2, 0), (3, 7), (4, 8), (2, 9)}

(C) {(1, 3), (2, 5), (3, 7), (4, 9), (5, 11)}

(D) {(3, 9), (4, 16), (5, 25), (6, 36), (7, 35)}

7. Neel has Y cars. He shares $\frac{1}{20}$ of his cars to his brother Ken.

How many cars does Neel have now ?

(A) 19Y

(B) $\frac{19Y}{20}$

(C) 20Y

(D) Y

8. Sai drives 205 miles in G hours. What is the average speed of the car ?

(A) $\frac{205}{G}$

(B) 205

(C) $\frac{G}{205}$

(D) G

9. If the length of a square K is reduced by half, then by what percent of its perimeter is decreased by to its original?

(A) 50% (B) 25%

(C) 60% (D) 75%

10. Mr. K bought four equally priced pizzas and a $3 bag of potato chips. If he spends $39, find the cost of each pizza and the equation.

(A) $4x-3=39;\ x = \frac{21}{2}$ (B) $3x+4=39;\ x = \frac{35}{3}$

(C) $4x+3=39;\ x = 9$ (D) $4x+3=39;\ x = \frac{21}{2}$

11. Write the algebraic expression of the statement below:

5 more than a number is equal to 30

(A) $n^5 \leq 30$ (B) $5 - n = 30$

(C) $n + 5 = 30$ (D) $5n = 30$

12. Write the algebraic expression of the statement below:

The quotient of a number and 5 is equal to 34

(A) $5 - n = 34$ (B) $n - 5 < 34$

(C) $\frac{5}{n} < 24$ (D) $\frac{n}{5} = 34$

13. Working together, Scott and Andy stacked 1200 books in library for 4.8 hours. It takes Andy 8 hours to stack the 1200 books alone. In how many hours can Scott Stack the 1200 books alone?

(A) 19 hours

(B) 14 hours

(C) 15 hours

(D) 12 hours

14. Find the missing value from the below table:

IN PUT	OUTPUT
5	10
7	16
11	28
6	?

(A) 3

(B) 27

(C) 17

(D) 13

15. Find the correlation between the data given below

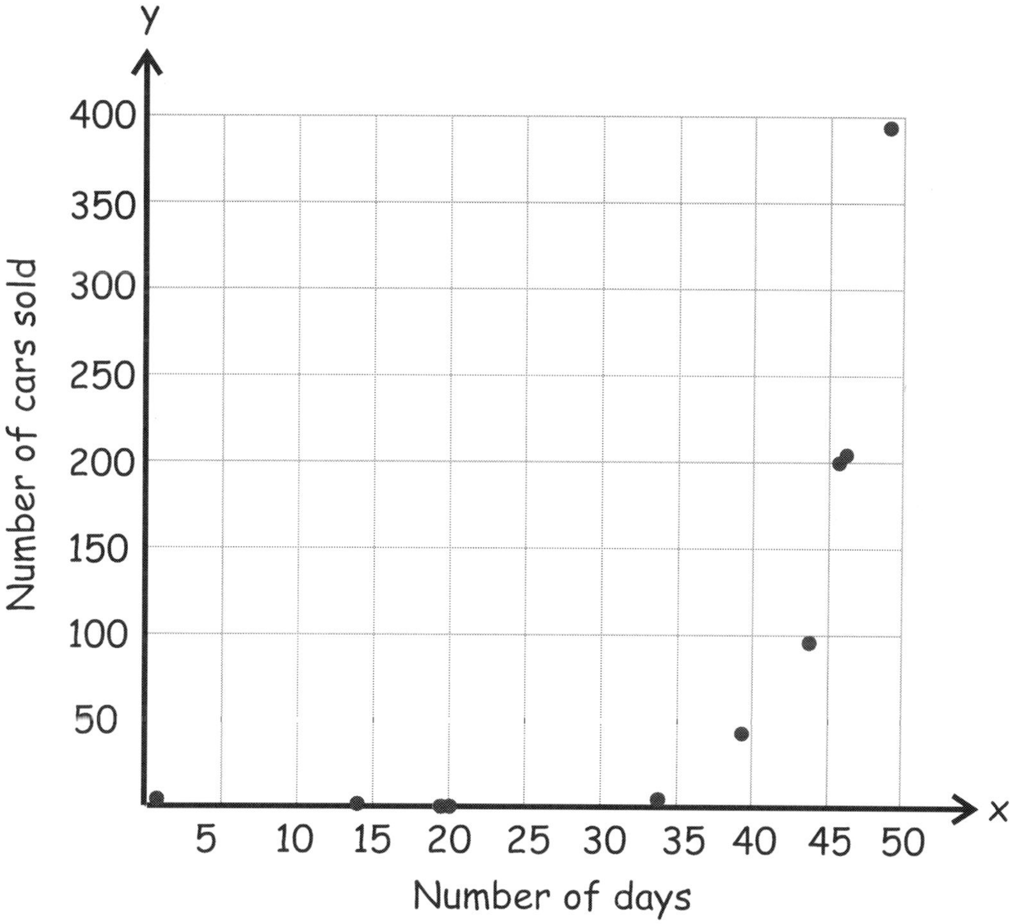

(A) Negative correlation

(B) No correlation

(C) Positive correlation

(D) None of these

SECTION D

Grace solved a number of puzzles this week as per the below dot plot. Based on this data, answer the questions 1 - 4

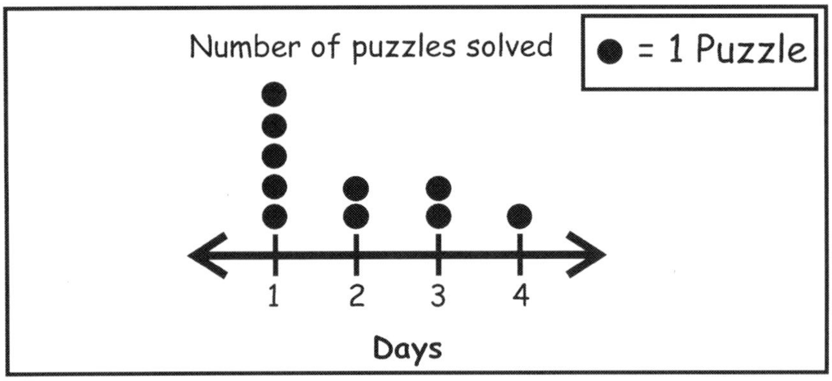

1. What is the median of the above data ?

(A) 2.0 (B) 2.5

(C) 3.0 (D) 4.0

2. What is the mean value of the above data ?

(A) 1.25 (B) 2.05

(C) 2.5 (D) 3.5

3. What is the range of the above data ?

(A) 6 (B) 3

(C) 5 (D) 4

4. What is the value of the mode from the above data ?

(A) 1 (B) 2

(C) 5 (D) 4

Tanya is designing a robot and verifies it a number of times. Number of trials on each day are recorded as below. Based on the below data answer the questions 5 - 8

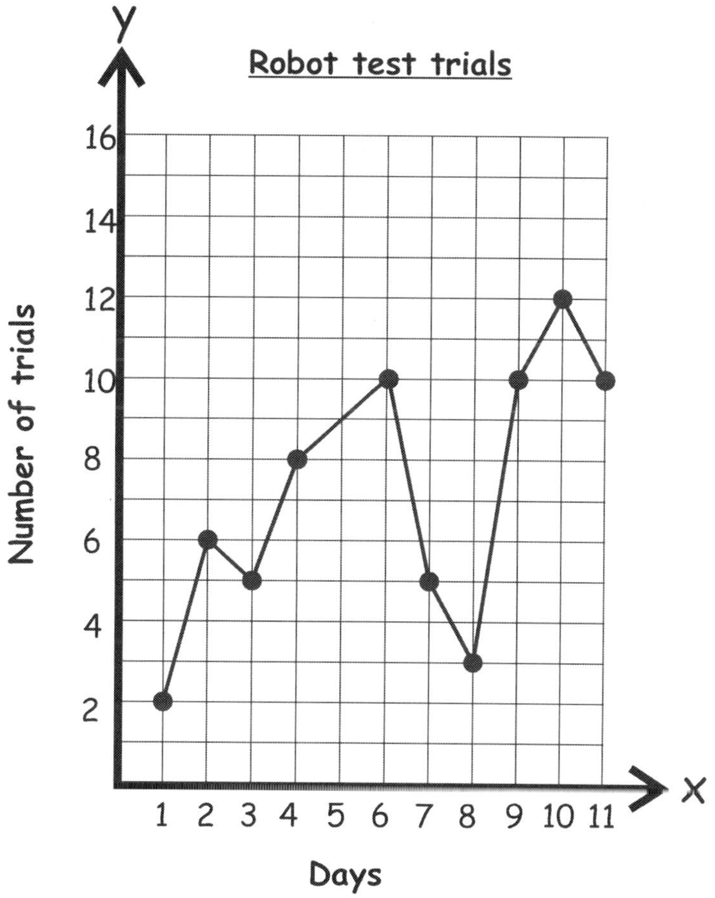

5. What is the median of the above data ?

(A) Median = 7
(B) Median = 10
(C) Median = 6
(D) Median = 7.5

6. What is the Mean of the above data ?

(A) Mean = 9.17
(B) Mean = 5.57
(C) Mean = 10.25
(D) Mean = 7.1

7.

(a) What is the Range of the above data ?

(A) 10 (B) 5

(C) 7 (D) 9

(b) What is the Mode of the above data ?

(A) 12 (B) 10

(C) 5 (D) 8

8. (a) Between which two days least number of trials are done by Tanya ?

(A) Days 8 and 9 (B) Days 2 and 3

(C) Days 1 and 2 (D) Days 4 and 6

(b) Between which two days greatest number of trials are done by Tanya ?

(A) Days 8 and 9 (B) Days 9 and 10

(C) Days 10 and 11 (D) Days 6 and 9

Daniel solved a number of physics problems over last 10 days. Below graph shows the number of problems solved by her. Based on this data, answer the questions 9 - 12.

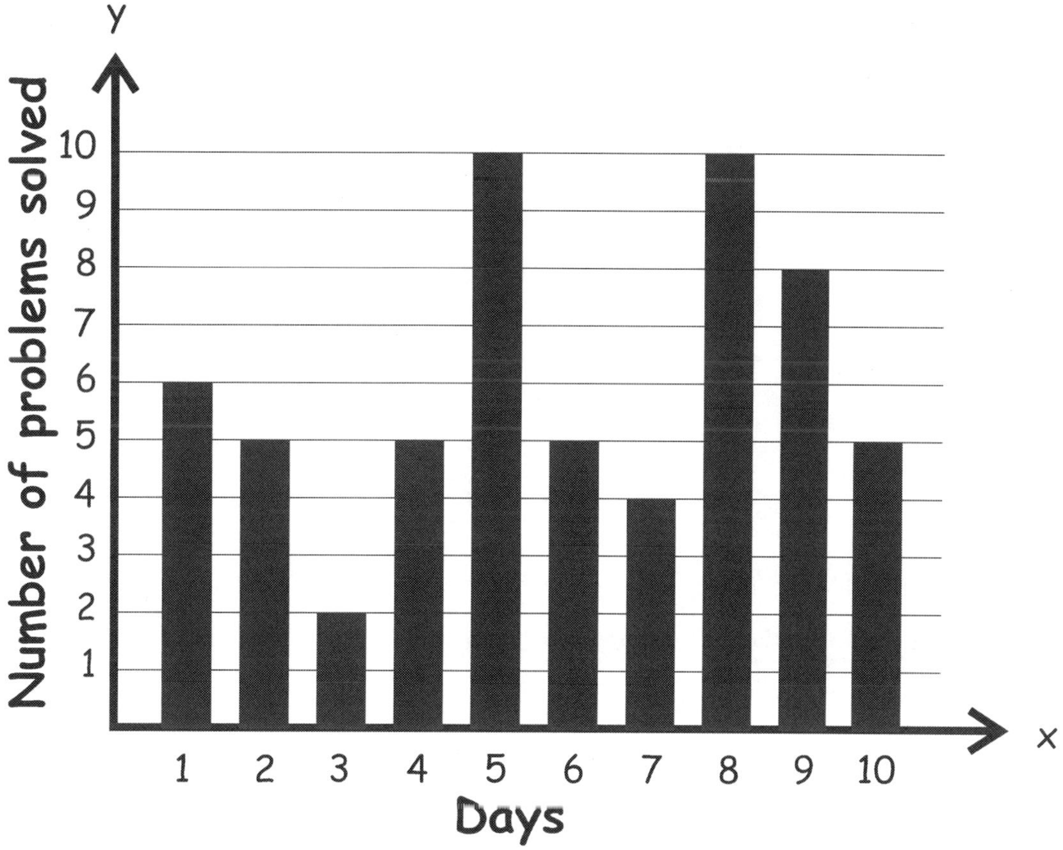

9. What is the median of the above data ?

(A) Median = 5

(B) Median = 2

(C) Median = 10

(D) Median = 6

10. What is the Mean of the above data ?

(A) Mean = 4.2

(B) Mean = 17.5

(C) Mean = 5.61

(D) Mean = 6.0

11. (a) What is the Range of the above data ?

(A) 10 (B) 7

(C) 8 (D) 2

(b) What is the Mode of the above data ?

(A) 6 (B) 8

(C) 10 (D) 5

12. (a) Between which two days least number of physics problems are solved by Daniel ?

(A) Days 6 and 7 (B) Days 4 and 6

(C) Days 5 and 6 (D) Days 9 and 10

(b) Between which two days greatest number of physics problems are solved by Daniel ?

(A) Days 1 and 2 (B) Days 3 and 4

(C) Days 4 and 5 (D) Days 4 and 6

13. A cocktail juice (C) was made by mixing 8 liters of juice A and 2 liters of juice B. Juice A contains 51% of Orange juice and juice B contains 61% of orange juice. Cocktail juice (C) contains what percentage of orange juice ?

(A) 53% (B) 62%

(C) 51% (D) 55%

14. Find the slope of the straight line from the graph given below.

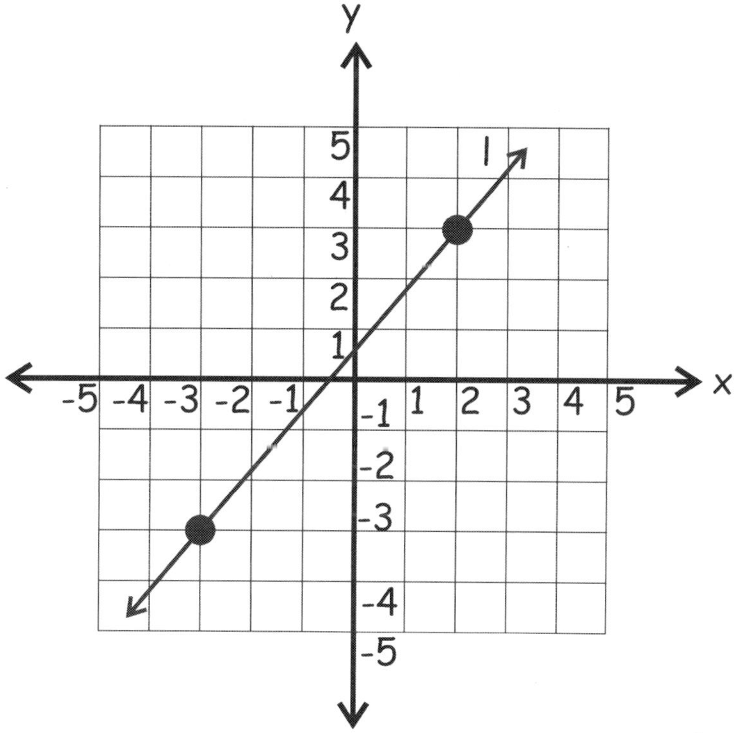

(A) -1 (B) $\frac{6}{5}$

(C) 1 (D) $\frac{1}{4}$

15. Find the missing $\angle x$ from the below figure ?

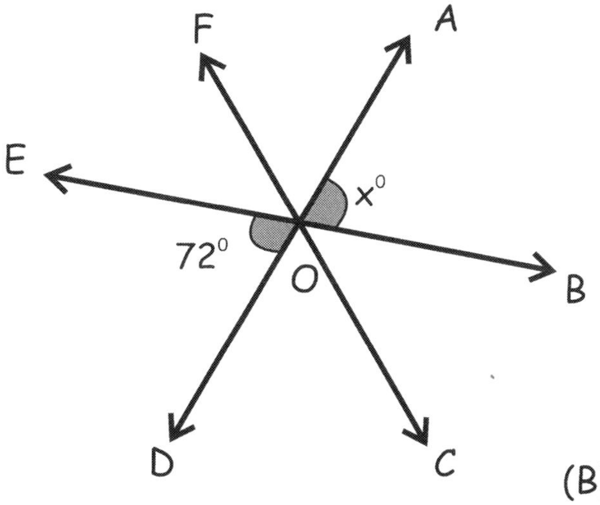

(A) 72^0 (B) 140^0

(C) 108^0 (D) 296^0

IAAT
PRACTICE TEST
TEST - 8

SECTION A

1. Ishaan has 36 yards of a wire. How many pieces of $\frac{6}{7}$ can he cut?

(A) 31

(B) 45

(C) 32

(D) 42

2. A train of length 54 meters has 10 cars equally spaced. What is the distance between 2nd and 6th car ?

(A) 21

(B) 27

(C) 24

(D) 26

3. Evaluate $(75) + (-74) + (-1) - 0 = \boxed{?}$

(A) 0

(B) 1

(C) -1

(D) 2

4. Evaluate $(200.5)(0.65) = \boxed{?}$

(A) 129.325

(B) 130.033

(C) 130.325

(D) 130.335

5. What is the Least Common Multiple (LCM) of below numbers

25 and 100

(A) 25

(B) 100

(C) 4

(D) 5

6. Ivana and Ruth scored a total of twenty two points in their last basketball game. Ruth scored two more points than Ivana. How many points did Ivana score ?

(A) $2x - 2 = 22$; $x=12$

(B) $x + 2 = 22$; $x=20$

(C) $x + 2 = 22$; $x=12$

(D) $2x + 2 = 22$; $x=10$

7. Jim reads at an average of 2 books per day over summer break of 53 days. How many books did he read in total ?

(A) 106

(B) 55

(C) 102

(D) 104

8. Ritvik solves 1 problem for every 8 sec. How many problems will he solve in 4 minutes ?

(A) 2

(B) 60

(C) 12

(D) 30

9. If Ram has 10 more marbles than Shyam. Shyam has 15 more marbles than Gopal who has 20 marbles. How many marbles are there all together ?

(A) 100 (B) 80

(C) 75 (D) 105

10. If the length of the rectangle is 20% more than width. If the width of the rectangle is 10cm. What is the perimeter ?

(A) 44 cm (B) 22 cm

(C) 120 sq cm (D) 12 cm

11. Tejas tosses 2 coins. What is the probability of having a two heads ?

(A) 4 (B) $\frac{1}{2}$

(C) $\frac{1}{8}$ (D) $\frac{1}{4}$

12. Express the below number in scientific notation.

0.000817

(A) 8.17×10^{4} (B) 8.17×10^{-4}

(C) 81.7×10^{5} (D) 81.7×10^{-3}

13. Solve the below inequality

$$\frac{5+x}{10} \leq 1$$

(A) $x \leq -18$

(B) $x \geq -1$

(C) $x \leq 5$

(D) $x \leq -1$

14. A model statue is 4 in tall. If it was built with a scale of 1 in : 2.8 ft, how tall is the real statue?

(A) 1.4 ft

(B) 11.2 ft

(C) 0.9 ft

(D) 6.8 ft

15. Write the below percent as a fraction.

23.5%

(A) 946

(B) $\frac{47}{200}$

(C) $94\frac{2}{6}$

(D) $23\frac{1}{3}$

SECTION B

1. Evaluate 3H, where $\frac{H}{2} = 25.$

(A) 50

(B) 75

(C) 25

(D) 150

2. Circumference of a circle "C" with radius 7 m is ?

(A) 14 π m

(B) π m

(C) 14 m

(D) 7 m

3. Evaluate K where $21K = 4444 - K.$

(A) K = 22

(B) K = 220

(C) K = 202

(D) K = 20

4. Solve for U and T where U and T are positive integers.

$U + T = 16$ and $U - T = 4$

(A) -10 and 6

(B) 10 and -6

(C) 12 and 4

(D) 10 and 6

5. If $R = 7$ and $P = 2R$ then $RP = ?$

(A) 49

(B) 14

(C) 2

(D) 98

6. Evaluate $(9T \times T) + (U \times U)$ where $T = -2$ and $U = -5$.

(A) -61

(B) 61

(C) 11

(D) -11

7. If $K\Delta P = 2K \times K - P \times P$. then what is the value of

$$(4\Delta 2) + (5\Delta 1) \boxed{?}$$

(A) 79

(B) 76

(C) 77

(D) 71

8. Find the value of the below

24% of 5 cm

(A) 44.9 cm

(B) 1.2 cm

(C) 3560 cm

(D) 20.8 cm

9. If U = -7 and T = -11 then

$$U - T = \boxed{?}$$

(A) -4　　(B) 18

(C) 4　　(D) -18

10. $U = \frac{1}{2}$ and 2U - 7T = -13 then what is the value of T ?

(A) -2　　(B) $\frac{-12}{7}$

(C) $\frac{12}{7}$　　(D) 2

11. Ram can stack 100 books in piles of 20 in 20 minutes.
He stacks B books in 2 hours. Using ration and proportion which of the below statements are true.

(A) $\frac{20}{100} = \frac{B}{120}$　　(B) $\frac{100}{20} = \frac{B}{120}$

(C) $\frac{100}{20} = \frac{120}{B}$　　(D) 2000 = 120 B

12. Express the below algebraic expression in words:

$$\boxed{n - 6 = 16}$$

(A) 6 divided by a number is equal to 16
(B) A number decreased by 6 is equal to 16
(C) 6 increased by a number is equal to 16
(D) 6 decreased by a number is equal to 16

13. Evaluate $5H^2 - 7H + 12$ where $H^{101} = -1$.

(A) 24

(B) 14

(C) -14

(D) 22

14. Simplify $-2(10r + 1) - 2$

(A) 2r - 10

(B) 2r - 19

(C) -20r - 4

(D) -20r + 6

15. Simplify $-170 \geq -17r$

(A) $r \geq 187$

(B) $r \leq -153$

(C) $r \geq 2890$

(D) $r \geq 10$

SECTION C

1. Sam has a food stand at a local fair. For every eight cheese sandwiches he sells five chicken sandwiches. If he has sold 50 chicken sandwiches in last four hours then how many cheese sandwiches did he sold?

(A) 80 (B) 16

(C) 88 (D) 78

2. If T = 5U - 5 and U = 12, what is the value of T ?

(A) 65 (B) 50

(C) 55 (D) 75

3. The ratio of rose plants to evergreen plants at a local nursery are 9:2. Tf there are 8 evergreen plants then, how many rose plants are there?

(A) 30 (B) 27

(C) 42 (D) 36

4. Evaluate $\frac{13}{7} \div 10$

(A) $\frac{3}{70}$ (B) $\frac{1}{70}$

(C) $\frac{13}{70}$ (D) $\frac{13}{17}$

5. Evaluate $126 = 5(6 - 7k) + 3k$

(A) 7

(B) -3

(C) -5

(D) 8

6. Which of the below is not function ?

(A) {(0, 0), (1, 1), (2, 2), (5, 1), (5, 5)}
(B) {(2, 6), (3, 7), (4, 8), (5, 9), (6, 10)}
(C) {(1, 3), (2, 5), (3, 7), (4, 9), (5, 11)}
(D) {(1, 5), (2, 6), (3, 7), (4, 8) ,(5, 9)}

7. 21 pieces of brownies need to be split among X friends. What is the share of each ?

(A) $\frac{21}{X}$

(B) 21 X

(C) $\frac{X}{21}$

(D) 21 + X

8. Sam bought P pounds of Blue berries and then got Y pounds of Cherries and then got 5 pounds of apples. How many pounds of fruits he bought?

(A) P + Y

(B) 5PY

(C) P + Y - 5

(D) P + Y + 5

9. Average two numbers is 24. A number is then added to the group to get an average of 30. What is the third number?

(A) 41 (B) 39

(C) 30 (D) 42

10. Evaluate $7(-3n + 6) = 126$

(A) -8 (B) 9

(C) -4 (D) 12

11. Write the algebraic expression of the statement below:

The difference of p and 22 is equal to 34

(A) $22(p) = 34$ (B) $22^{3} \geq 34$

(C) $p - 22 = 34$ (D) $\frac{p}{22} \leq 34$

12. Write the algebraic expression of the statement below:

13 less than k is equal to 33

(A) $\frac{k}{2} = 33$ (B) $k - 13 = 33$

(C) $2k = 33$ (D) $k + 13 = 33$

13. Don said to Sam, "500 reduced by twice my age is 326." What is Don's age?

(A) 78
(B) 87
(C) 92
(D) 88

14. Find the missing value from the table below:

IN PUT	OUTPUT
2	2
4	6
7	?
8	14
11	20

(A) 14
(B) 12
(C) 8
(D) 10

15. Find the correlation between the data given below

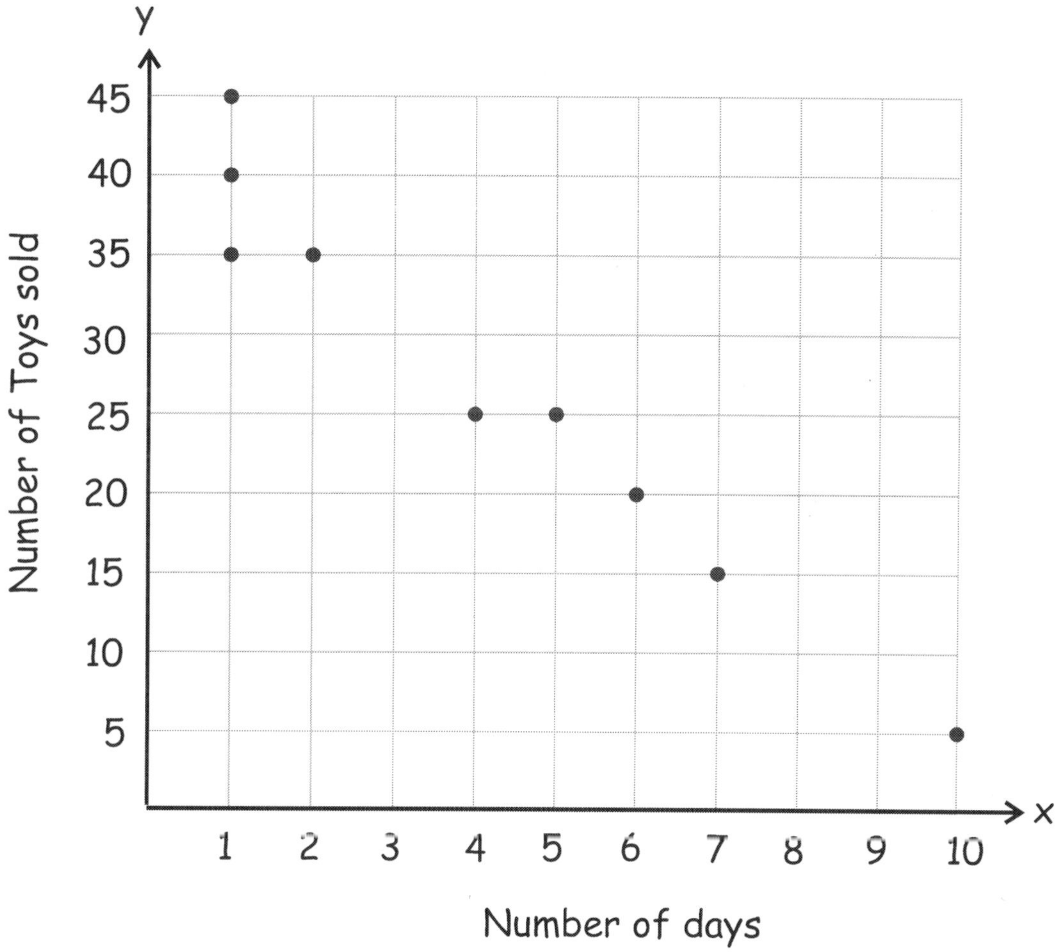

(A) No correlation

(B) Negative correlation

(C) Positive correlation

(D) None of these

SECTION D

Grade 7 students performance from the math quiz is shown below. Based on this data answer the questions 1-4

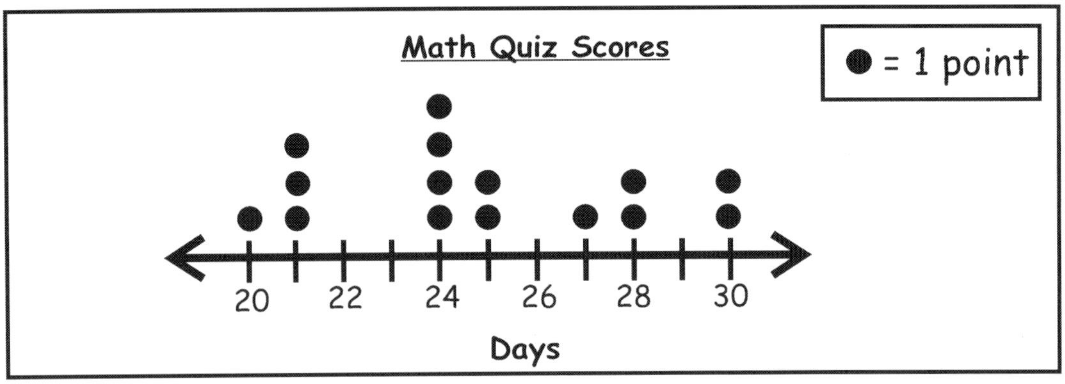

1. What is the median of the above data ?

(A) 24 (B) 25

(C) 23 (D) 30

2. What is the mean value of the above data ?

(A) 20.13 (B) 22.51

(C) 19.23 (D) 22.13

3. What is the range of the above data ?

(A) 28 (B) 16

(C) 10 (D) 12

4. What is the value of the mode from the above data ?

(A) 23 (B) 26

(C) 25 (D) 24

Beautiful carpet store charges per square feet of each carpet on sale . Based on the below data answer the questions 5 - 8.

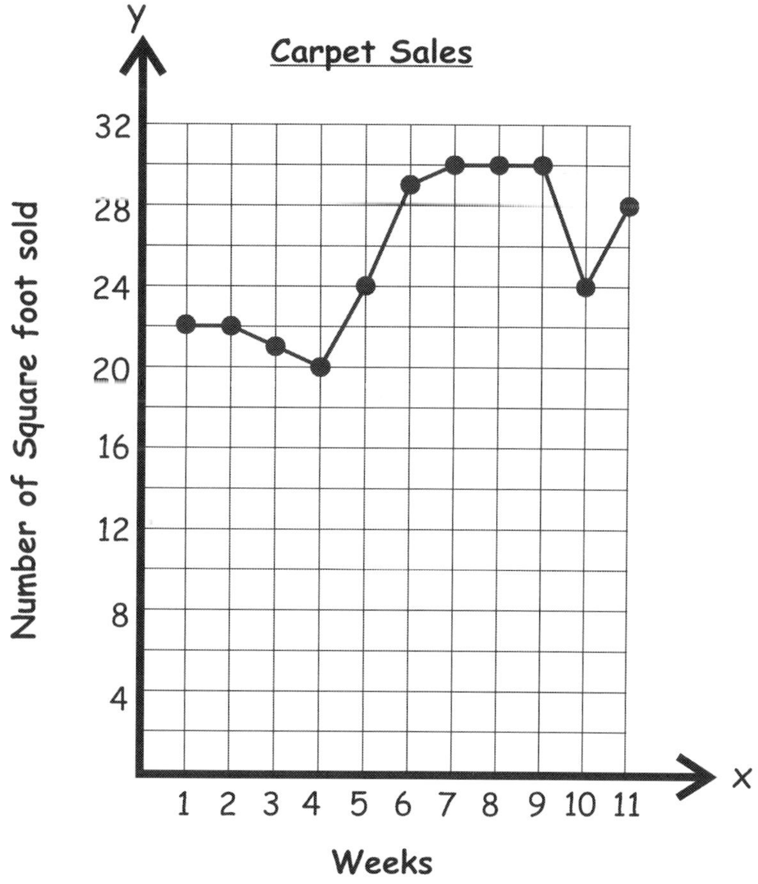

5. What is the median of the above data ?

(A) Median = 24

(B) Median = 28

(C) Median = 20

(D) Median = 21

6. What is the Mean of the above data ?

(A) Mean = 25.44

(B) Mean = 25.54

(C) Mean = 25.46

(D) Mean = 25.06

7. (a) What is the Range of the above data ?

(A) 10 (B) 7

(C) 5 (D) 9

(b) What is the Mode of the above data ?

(A) 22 (B) 24

(C) 20 (D) 30

8. (a) Between which two days least number of sales are done ?

(A) Days 5 and 6 (B) Days 1 and 2

(C) Days 10 and 11 (D) Days 4 and 5

(b) Between which two days highest number of sales are done ?

(A) Days 7 and 8 (B) Days 10 and 11

(C) Days 4 and 5 (D) Days 5 and 6

PWPS Rocket club test launches their rockets as a preperation for the upcoming competition. The test launches are recorded as below. Based on this data, answer the questions 9 - 12.

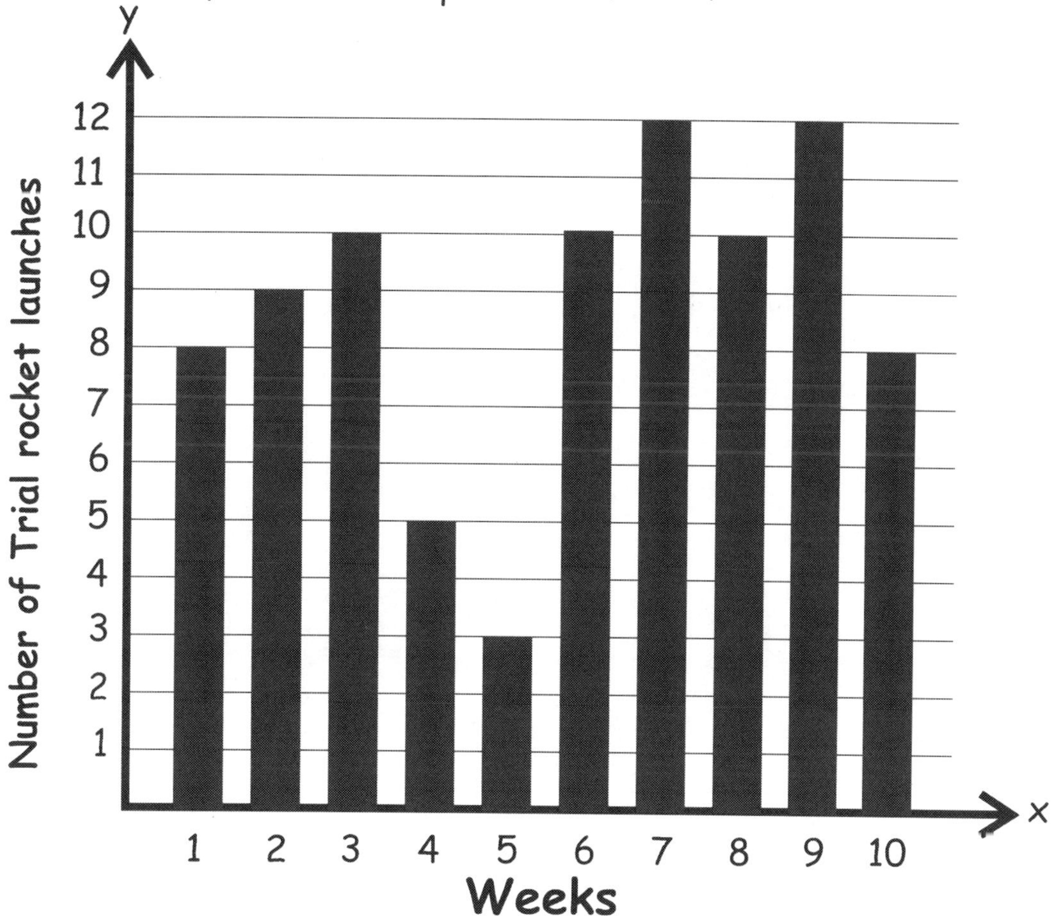

9. What is the median of the above data ?

(A) Median = 8.5

(B) Median = 11

(C) Median = 9.5

(D) Median = 9

10. What is the Mean of the above data ?

(A) Mean = 8.7

(B) Mean = 8.4

(C) Mean = 6.87

(D) Mean = 8.17

11. (a) What is the Range of the above data ?

(A) 9 (B) 4

(C) 8 (D) 11

(b) What is the Mode of the above data ?

(A) 12 (B) 10

(C) 8 (D) 3

12. (a) Between which two days least number of trials are done ?

(A) Days 8 and 9 (B) Days 10 and 1

(C) Days 7 and 8 (D) Days 3 and 4

(b) Between which two days highest number of trials are done ?

(A) Days 2 and 3 (B) Days 7 and 9

(C) Days 5 and 9 (D) Days 6 and 3

13. A two liters of Orange and Apple juice contains 45% of Orange juice. When a three liters of soda is mixed to it , What percentage of Orange juice is in this mixture ?

(A) 15% (B) 32%

(C) 25% (D) 18%

14. Find the slope of the straight line from the graph given below.

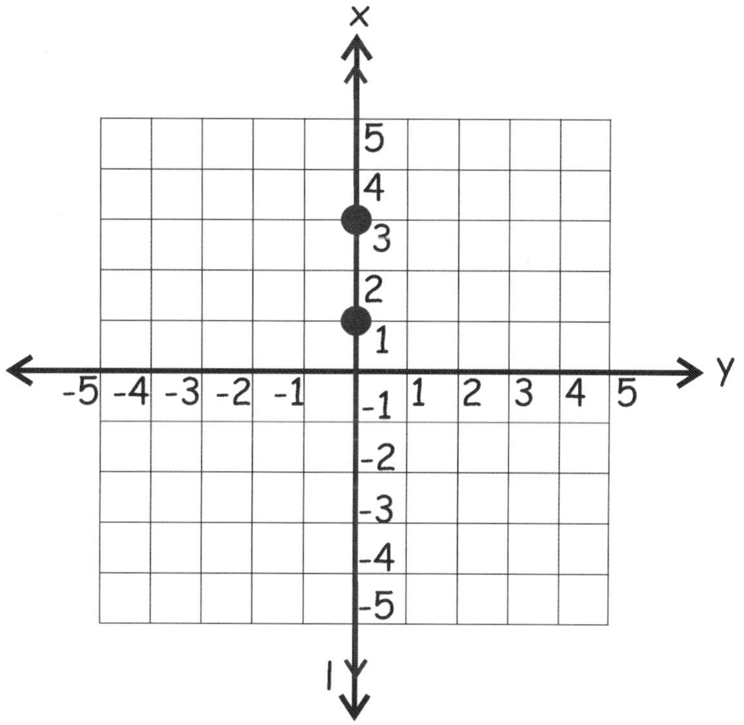

(A) $\frac{1}{6}$ (B) undefined

(C) $\frac{-1}{11}$ (D) $\frac{1}{2}$

15. Find the missing $\lfloor x$ from the below figure ?

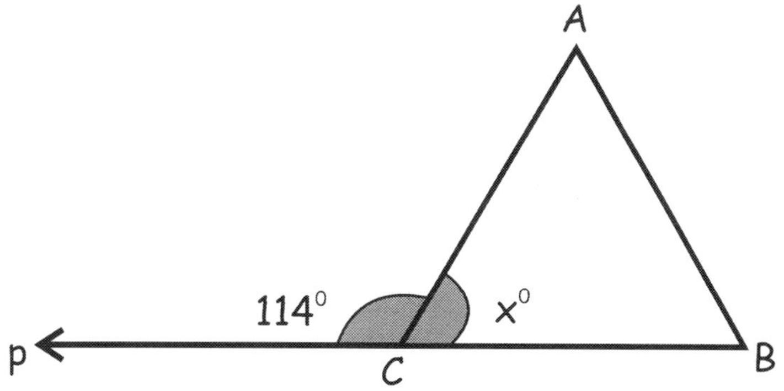

(A) 66° (B) 114°

(C) 146° (D) 216°

IAAT
PRACTICE TEST
TEST - 9

SECTION A

1. Tejas wants to divide a 16 yards of cloth into pieces of $\frac{2}{11}$ yard. How many such pieces can he cut ?

(A) 32 (B) 88

(C) 44 (D) 11

2. Sid is selling fruit baskets. He sells 121 baskets with 1331 fruits in total. If each basket has equal number of fruits, how many fruits does each basket have ?

(A) 11 (B) 1

(C) 121 (D) 21

3. Evaluate $(-35) + (10) + \frac{27}{3} - \frac{10}{2} = \boxed{?}$

(A) 21 (B) 19

(C) -17 (D) -21

4. Evaluate $12.12 \times 1.2 = \boxed{?}$

(A) 44.544 (B) 14.544

(C) 14.444 (D) 14.454

5. Evaluate $168 = 7(3x + 3)$

(A) 7

(B) 11

(C) -7

(D) -16

6. What is the Greatest Common Factor of the below numbers

110, 5, 20

(A) 10

(B) 5

(C) 2

(D) 4

7. Gary runs at an average speed of 3 miles per half hour.
If she ran 30 miles what is the total time she took ?

(A) 3

(B) 5

(C) 15

(D) 10

8. Ria solves 8 out of 9 problems correctly on an average.
If she solves 108 problems in a day, how many problems were wrong ?

(A) 60

(B) 81

(C) 45

(D) 96

9. Kiran buys 4 bottles of 20 oz cooking oil for $12. What is the price per Ounce ?

(A) 0.15 (B) 0.11

(C) 0.18 (D) 0.20

10. If the area of the square is 121 units, then what is its perimeter ?

(A) 44 (B) 40

(C) 22 (D) 121

11. Rik tosses 2 coins. What is the probability of having a one head and a tail ?

(A) $\frac{1}{2}$ (B) 4

(C) $\frac{1}{4}$ (D) 2

12. Evaluate the below

$$(1.7 \times 10^{2})(8 \times 10^{3}) = \boxed{?}$$

(A) 0.136×10^{0} (B) 1.36×10^{0}

(C) 2.125×10^{-4} (D) 21.25×10^{-4}

13. Solve the below inequality.

$$0 \geq -3 + \frac{m}{5}$$

(A) $m \leq -22$

(B) $m \leq 15$

(C) $m \leq -28$

(D) $m \leq 6$

14. Allison baked a cake for Henry's birthday; $\frac{4}{7}$ of the cake was eaten at the birthday party. The next day, Henry ate half of what was left. How much of the cake did Henry ate the next day ?

(A) $\frac{1}{7}$

(B) $\frac{1}{4}$

(C) $\frac{2}{7}$

(D) $\frac{3}{14}$

15. If the original price of an oil change is $29.95 and the discount offered on that is 50%, then what is the sale price of it ?

(A) $14.98

(B) $16.99

(C) $14.99

(D) $14.58

SECTION B

1. The ratio of rulers to erasers in a box is 4:3. If there are 12 rulers then, how many erasers are there ?

(A) 12 (B) 9

(C) 11 (D) 16

2. Simplify

$$3\sqrt{24} + 3\sqrt{54} = \boxed{?}$$

(A) $15\sqrt{6}$ (B) $36\sqrt{6}$

(C) $27\sqrt{6}$ (D) $21\sqrt{6}$

3. Evaluate j - h + 5 where h = 5, and j = 3

(A) 1 (B) 2

(C) 7 (D) 4

4. Solve for U and T, where U and T are positive integers.

$$U - T = 4 \text{ and } U + T = 36$$

(A) 20 and 18 (B) -18 and 18

(C) 20 and 17 (D) 20 and 16

5. If $R \times R = 36$ what is the value of R ?

(A) 18
(B) 6
(C) 2
(D) 9

6. Evaluate $\boxed{U \times U - 2T}$ where $T = 5$ and $U = -5$.

(A) 15
(B) 10
(C) 25
(D) 13

7. If $K\#P = K \times K - P$, what is the value of $\boxed{(4\#2) + (3\#5)}$

(A) 20
(B) 19
(C) 18
(D) 17

8. Write the algebraic expression of the below:

The quotient of a number and 6 is equal to 43

(A) $\frac{n}{6}=43$
(B) $n^2 = 43$
(C) $n - 6 = 43$
(D) $\frac{6}{n}=43$

9. If $U = 33$ and $T = -22$ then $\boxed{\frac{U-T}{5} = \boxed{?}}$

(A) 55
(B) $\frac{11}{5}$
(C) 11
(D) $\frac{5}{11}$

10. $U = \frac{1}{4}$ and $4U - 2T = 11$ then what is the value of T ?

(A) -5

(B) 11

(C) 5

(D) 22

11. Solve for 'P' from the below proportion.

$$\frac{8}{P} = \frac{2}{6}$$

(A) 1

(B) 2

(C) 4

(D) 24

12. Express the below algebraic expression in words.

$$\frac{a}{7} = 39$$

(A) The quotient of 7 and a is equal to 39

(B) The quotient of a and 7 is equal to 39

(C) 7 to the a is equal to 39

(D) The sum of a and 7 is equal to 39

13. 5% of how many inches is 48 inches ?

(A) 2.4 inches
(B) 19.8 inches
(C) 960 inches
(D) 240 inches

14. Simplify $-x - 9(-x + 6)$

(A) 32 + 55x
(B) 8x - 54
(C) 40 + 56x
(D) 40 + 55x

15. Simplify $b - (-17) < 36$

(A) $b < -53$
(B) $b > -612$
(C) $b < -612$
(D) $b < 19$

SECTION C

1. Zara is preparing for her upcoming exams. Every day she studies history and math in the ratio of 2 to 9, This week she studies six hours of history then how many hours of math did she study ?

(A) 22 (B) 27

(C) 23 (D) 29

2. If $\frac{T}{5}=\frac{1}{3}U+\frac{7}{5}$ and U = 6, then what is the value of T ?

(A) 8 (B) -17

(C) 13 (D) 17

3. Peter has ten white shirts for every two blue shirts in his collection. He now has 30 white shirts then how many shirts does he has now ?

(A) 6 (B) 2

(C) 11 (D) 9

4. The sum of eleven and seven times a number is -17. Find the number and the expression

(A) $7 + y + 11=-17$; $y=-1$ (B) $7y + 11=-17$; $y=-4$

(C) $7y + 11=-17$; $y= \frac{-5}{7}$ (D) $7y - 11=-17$; $y=\frac{-6}{7}$

5. Stacy lives on a cliff. She lives 652 feet above sea level. When she travels to town, she travels down 491 feet. What is the elevation of the town ?

(A) 261 ft above sea level

(B) 161 ft below sea level

(C) 141 ft above sea level

(D) 161 ft above sea level

6. Which of the below is not a function ?

(A) {(-1, 0), (0, 1), (1, 2), (2, 3), (3, 4)}

(B) {(2, 3), (4, 15), (5, 24), (6, 35), (7, 48)}

(C) {(1, 3), (2, 5), (3, 7), (3, 9), (5, 11)}

(D) {(3, 0), (5, 7), (7, 9), (9, 11), (11, 13)}

7. One stack has 15 books and there are Y such stacks and 5 individual books. How many books are there in total ?

(A) 15Y - 5

(B) 15Y + 5

(C) 3Y

(D) 5Y

8. The number of baskets Rik made in the soccer game are one less than triple number of the goals made by Ryan. Rayan makes 2 goals. How many goals were made my Rik ?

(A) 5

(B) 1

(C) 6

(D) 7

9. Average rate of work of Sam and Shyam is 45 hours this week. If Sam works 50 hours per week, how many hours did Shyam work this week ?

(A) 50 (B) 90

(C) 450 (D) 45

10. Mt. Everest, the highest elevation in Asia, 29,028 feet above sea level. The dead sea, the lowest elevation, is 1,312 feet below sea level. What is is the difference between these two elevations ?

(A) 27,716 ft (B) 30,340 ft

(C) 31,432 ft (D) 21,776 ft

11. Write the algebraic expression of the statement below:

The a power of 13 is 17

(A) $a - 13 \leq 17$ (B) $a^{13} > 17$

(C) $13 + a > 17$ (D) $13^{a} = 17$

12. Write the algebraic expression of the statement below:

A number minus 11 is equal to 41

(A) $n \cdot 11 = 41$ (B) $11 - n > 41$

(C) $11 - n \leq 41$ (D) $n - 11 = 41$

13. Kim purchased a magnet for $6 and puzzles for $5 each. She spent a total of $26. How many puzzles did you buy ?

(A) 4 (B) 7

(C) 12 (D) 2

14. Find the missing value from the below table

IN PUT	OUTPUT
5	20
6	24
9	36
12	?
10	40

(A) 14 (B) 48

(C) 8 (D) 10

15. Find the correlation between the data given below

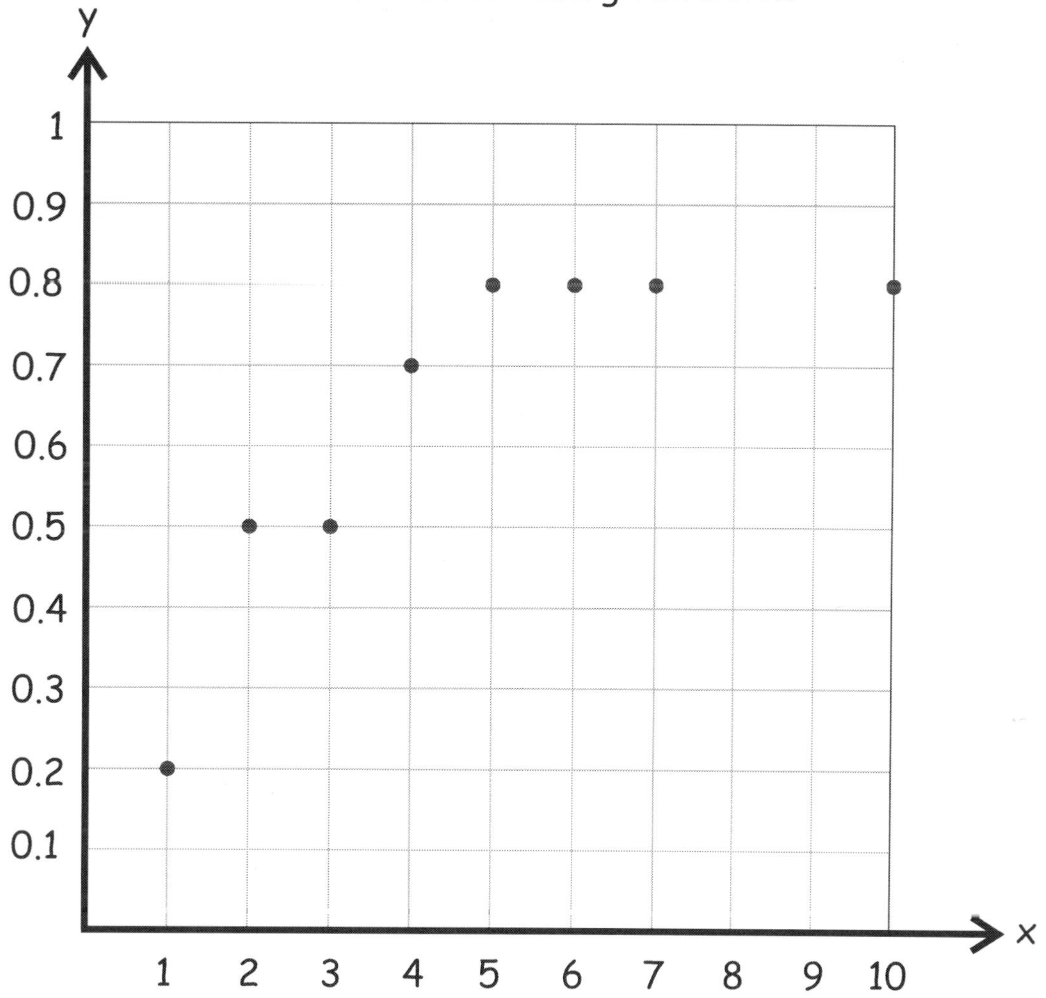

(A) No correlation

(B) Positive correlation

(C) Negative correlation

(D) None of these

SECTION D

Jane makes beautiful flower bouquets and sells them at a local fare. Her sales this week are shown below. Based on this data, answer the questions 1 - 4

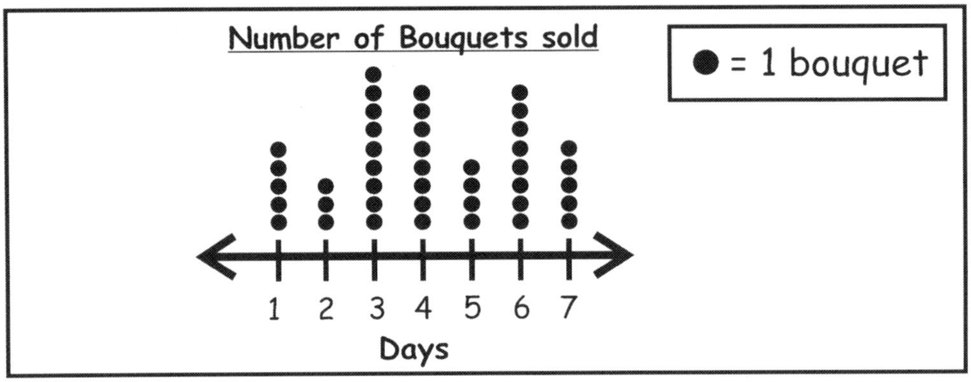

1. What is the median of the above data ?

(A) 5 (B) 3

(C) 7 (D) 9

2. What is the mean value of the above data ?

(A) 4 (B) 7.5

(C) 6 (D) 9.2

3. What is the range of the above data ?

(A) 3 (B) 8

(C) 5 (D) 6

4. What is the value of the mode from the above data ?

(A) 8 and 5 (B) 10

(C) 3 (D) 4

Miss Maria is analyzing the Geography Bee scores of grade six students and plotted the below graph. Based on the below data answer the questions 5 - 8

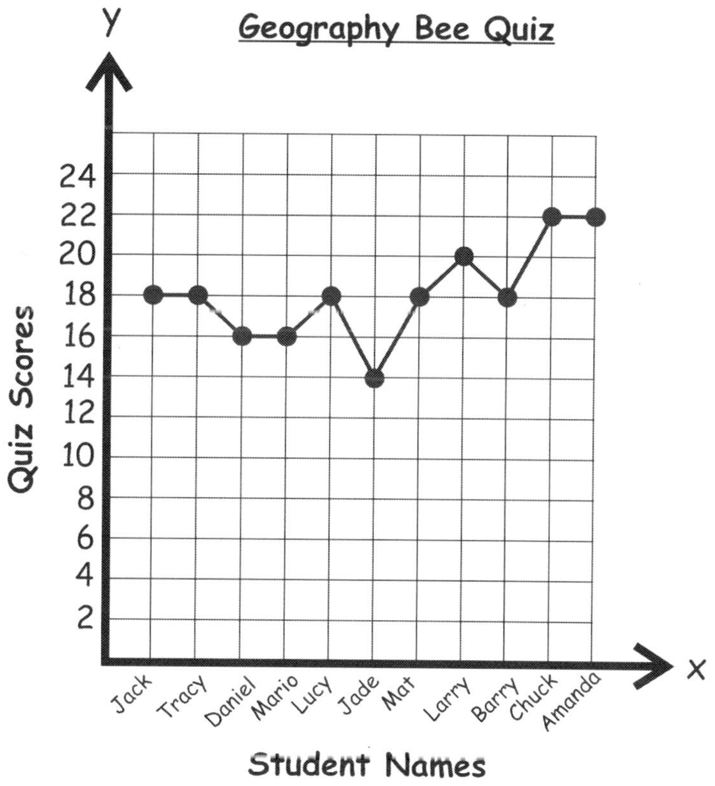

5. What is the median of the above data ?

(A) Median = 22 (B) Median = 18

(C) Median = 20 (D) Median = 16

6. What is the Mean of the above data ?

(A) Mean = 18.08 (B) Mean = 18.18

(C) Mean = 18.88 (D) Mean = 18.01

7. (a) What is the Range of the above data ?

(A) 2 (B) 10

(C) 6 (D) 8

(b) What is the Mode of the above data ?

(A) 18 (B) 14

(C) 10 (D) 22

8. (a) Which two students got the same scores ?

(A) Jack and Daniel (B) Jack and Tracy

(C) Tracy and chuck (D) Jade and larry

(b) Which two students scored the highest points ?

(A) Chuck and Amanda (B) Lucy and Mat

(C) Daniel and Maria (D) Jade and Barry

Peter solved a number of math problems over last 6 days. Below graph shows the number of problems solved by him. Based on this data, answer the questions 9 - 12.

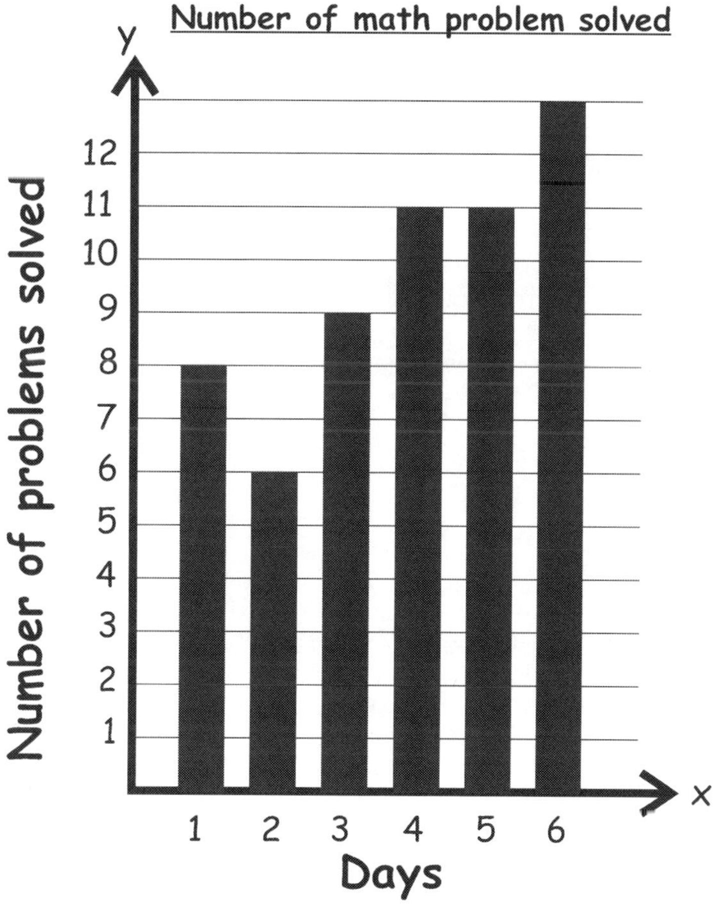

9. What is the median of the above data ?

(A) Median = 9
(B) Median = 13
(C) Median = 10
(D) Median = 6

10. What is the Mean of the above data ?

(A) Mean = 9.06
(B) Mean = 8.68
(C) Mean = 8.77
(D) Mean = 9.67

11. (a) What is the Range of the above data ?

(A) 10 (B) 7

(C) 15 (D) 9

(b) What is the Mode of the above data ?

(A) 9 (B) 5

(C) 6 (D) 11

12. (a) Between which two days least number of sales are done ?

(A) Days 4 and 5 (B) Days 2 and 6

(C) Days 1 and 3 (D) Days 3 and 5

(b) Which two days had the same sales ?

(A) Days 4 and 5 (B) Days 3 and 5

(C) Days 3 and 6 (D) Days 1 and 4

13. Jack and Lucy are selling big and small Carved candles. Jack sold nine small and six big art pieces for a total of $213. Lucy sold six small and one big art pieces for a total of $94. What is the cost of each of small and big art pieces ?

(A) big carved candle = $15 , small carved candle = $13
(B) big carved candle = $16 , small carved candle = $10
(C) big carved candle = $10 , small carved candle = $8
(D) big carved candle = $16 , small carved candle = $13

14. Find the slope of the straight line from the graph given below.

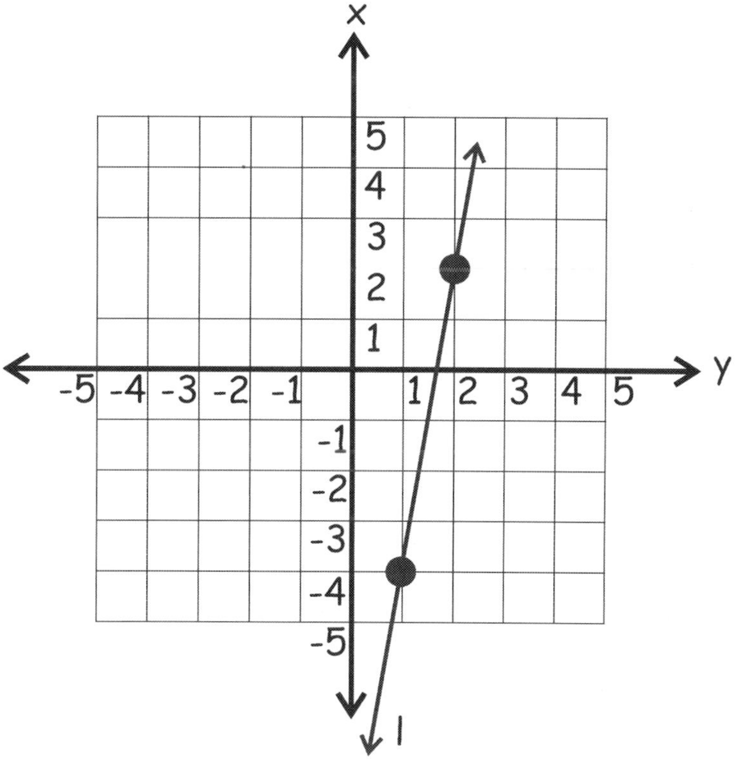

(A) undefined (B) $\frac{1}{6}$

(C) $\frac{-1}{6}$ (D) 6

15. Find the missing $\angle x$ from the below figure ?

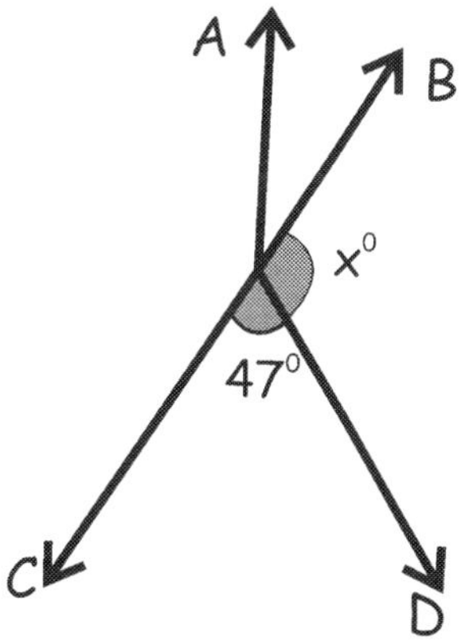

(A) 33° (B) 43°

(C) 133° (D) 123°

IAAT
PRACTICE TEST
TEST - 10
2
3
A
B
C

SECTION A

1. Rik earns each week \$200 in a part time job and spends $\frac{2}{5}$ of the money on video games and $\frac{1}{5}$ on clothes. How much money does he save in 8 weeks of summer part time job ?

(A) 40 (B) 120

(C) 640 (D) 150

2. Neha is making a string with Blue and Yellow beads. She uses 2 Blue beads for every Yellow bead. If she starts the string with Blue bead what will be color of 132nd beed in the string ?

(A) Yellow (B) White

(C) Blue (D) Green

3. Evaluate $(-5) + 8 \times \frac{5}{4} + 10 \times 2 - \frac{25}{5} = \boxed{?}$

(A) 5 (B) 45

(C) 25 (D) 20

4. Simplify $64.567 + 64.243 + 0.0005 + 3.0001 = \boxed{?}$

(A) 132.8160 (B) 131.0816

(C) 131.6018 (D) 131.8106

5. A photo is 20 in tall and 15 in wide. If it is reduced to a height of 4 in, then how wide will it be ?

(A) 6 in

(B) 3 in

(C) 4 in

(D) 5 in

6. The area of a triangle is the product of the base of the triangle times the height of the triangle, divided by two.

(A) $A = 2bh$

(B) $A = (bh)^2$

(C) $A = \frac{2h}{b}$

(D) $A = \frac{bh}{2}$

7. Anna drives the car at an average speed of 65 miles per hour and covers a distance of 195 miles. What is the total time he took to cover the distance ?

(A) 2 hours

(B) 3 hours

(C) 90 min

(D) 270 min

8. Tej is stacking books in the library at the rate of 5 min for each stack. Each stack has 20 books. How long does it take for him to arrange 2032 books in stacks ?

(A) 511 min

(B) 508 min

(C) 521 min

(D) 519 min

9. Sia buys 4 dozen cookies from the grocery store. If each dozen costs \$3 and there is sales tax of 5% to the bill. How much did she paid ?

(A) \$12.60

(B) \$3.05

(C) \$0.60

(D) \$3.10

10. Perimeter of an equilateral triangle whose height is 10 units is 18 units. What is the area of the triangle ?

(A) 90 sq. units

(B) 180 sq. units

(C) 30 sq. units

(D) 35 sq. units

11. What is the probability of drawing a queen in the deck of cards ?

(A) $\frac{1}{4}$

(B) $\frac{1}{52}$

(C) $\frac{1}{13}$

(D) $\frac{1}{26}$

12. Evaluate the below

$$(7 \times 10^{3})(5.5 \times 10^{-3}) = \boxed{?}$$

(A) 1.273×10^{0}

(B) 38.5×10^{7}

(C) 38.5×10^{-1}

(D) 3.85×10^{7}

13. Solve the below inequality.

$$\frac{b + 6}{21} < 1$$

(A) $b < -13$

(B) $b < -28$

(C) $b < 15$

(D) $b < -30$

14. Find the greatest common factor for the below:

16 & 52

(A) 5

(B) 4

(C) 8

(D) 32

15. If the original price of a tie is $35.00 and the discount offered on that is 5%, then what is the sale price of the tie ?

(A) $31.25

(B) $33.15

(C) $32.33

(D) $33.25

SECTION B

1. Simplify the below

$$\sqrt{20V^2}$$

(A) $3V\sqrt{7V}$

(B) $3V^2\sqrt{6}$

(C) $7\sqrt{6V}$

(D) $2V\sqrt{5}$

2. Rita has two black handbags for every three brown hand bags in her collection. She buys more bags this months. If she now has 81 brown bags then how many black bags does she has now ?

(A) 55

(B) 54

(C) 47

(D) 39

3. Evaluate $\frac{13}{7} \div \frac{7}{5}$

(A) $\frac{6}{9}$

(B) $\frac{5}{49}$

(C) $\frac{13}{7}$

(D) $\frac{65}{49}$

4. Solve for U and T, where U and T are positive integers.

$$U + T = 44 \text{ and } U - T = 4$$

(A) 24 and 20

(B) -24 and 20

(C) 20 and -24

(D) -24 and -20

5. If $K \times K + 1 = 82$ what is the value of K ?

(A) 81

(B) 27

(C) 1

(D) 9

6. Evaluate 6T - 7U where T = -3 and U = -1

(A) 11

(B) 18

(C) -11

(D) -18

7. If $K\#P = K \times \frac{K}{P}$, then what is the value of (5#1)#5

(A) 125

(B) 25

(C) 625

(D) 5

8. Write the algebraic expression of the below:

A number times 5 is equal to 20

(A) $5 - n = 20$

(B) $5^n \geq 20$

(C) $n \cdot 5 = 20$

(D) $5 + n = 20$

9. If U = 3 and T = -13, then $\boxed{\frac{2U-3T}{9} = \boxed{?}}$

(A) $\frac{11}{3}$

(B) 5

(C) -5

(D) $\frac{-11}{3}$

10. If U = 7 and -2U - 2T - 1 = 35, then what is the value of T ?

(A) 5

(B) 25

(C) -25

(D) 10

11. Solve for 'n' from the below proportion.

$$\boxed{\frac{4}{n} = \frac{5}{8}}$$

(A) 6.4

(B) 7.1

(C) 8

(D) 3.09

12. Express the below algebraic expression in words.

$$\boxed{n + 12 = 45}$$

(A) n increased by 12 is equal to 45

(B) n minus 12 is equal to 45

(C) Twice n is equal to 45

(D) 12 divided by n is equal to 45

13. 90% of how many hours is 45 hours ?

(A) 18.8 hours

(B) 2842 hours

(C) 50 hours

(D) 28.4 hours

14. Simplify the below

$-10v - 3(2 - v)$

(A) -7v - 10

(B) -10v - 10

(C) -7v -6

(D) 4v - 6

15. Simplify the below

$-3 \le -5 - x$

(A) $x \le -2$

(B) $x \ge 8$

(C) $x \le 2$

(D) $x \le 8$

SECTION C

1. Simplify 8(6a + 2) = -128

(A) -3

(B) -7

(C) -5

(D) 6

2. If $H = \frac{7M-1}{2}$ and M = 12, then what is the value of H ?

(A) 40.5

(B) 41.5

(C) 43

(D) 42.7

3. Evaluate $-0.8 \div -2.5$

(A) 0.23

(B) 0..3

(C) 0.44

(D) 0.32

4. Linus wants to buy ribbon to make bookmarks. One bookmark will be $12\frac{1}{2}$ inches long and the other two will be $7\frac{1}{4}$ inches long. How much ribbon should he buy ?

(A) 24 inches

(B) 27 inches

(C) $19\frac{3}{4}$ inches

(D) 19 inches

5. A rectangle is 3 in wide and 2 in tall. If it is enlarged to a width of 18 in, then how tall will it be ?

(A) 12 in

(B) 10 in

(C) 3 in

(D) 108 in

6. Which of the below is not a function ?

(A) {(1, 3), ((2, 5), (3, 7), (5, 7), (5, 11)}

(B) {(2, 3), (4, 15), (5, 24), (6, 35), (7, 48)}

(C) {(2, 5), (3, 10), (4, 17), (5, 26), (6, 39)}

(D) {(-1, 0), (0, 1), (1, 2), (2, 3), (3, 4)}

7. If each milk carton weighs 3.18 gallons and if Rik buys P such cartons, what is the total weight of milk in gallons ?

(A) 3P

(B) 2.18P

(C) 0.32P

(D) 3.18P

8. Sahi solve 5 less than twice the math problems solved by Sia. If Sia solves 12 problems, then how many problems did Sahi solve ?

(A) 29

(B) 19

(C) 17

(D) 24

9. Ritvik ran 10 miles of average distance over two weeks. If he ran 8 miles in first week, then how many miles did he ran second week ?

(A) 15 (B) 12

(C) 20 (D) 8

10. The ratio of girls to boys in a grade is 6 to 5. If there are 24 girls in the grade then how many students are there altogether ?

(A) 44 (B) 14

(C) 24 (D) 34

11. Write the algebraic expression of the statement below :

Half of a number is equal to 46

(A) $2n = 46$ (B) $\frac{2}{2} > 16$

(C) $n + 2 = 46$ (D) $\frac{n}{2} = 46$

12. Write the algebraic expression of the statement below :

6 more than a number is equal to 50

(A) $6^n = 50$ (B) $n + 6 = 50$

(C) $2n = 50$ (D) $6 + n \geq 50$

13. Tom sold half of his paintings and then bought five more. He now has 24. With how many paintings did he begin to sell ?

(A) 38 (B) 44

(C) 37 (D) 42

14. Find the missing value from the table below :

IN PUT	OUTPUT
3	18
5	30
7	?
9	54
10	60

(A) 44 (B) 31

(C) 42 (D) 48

15. Find the correlation between the data given below

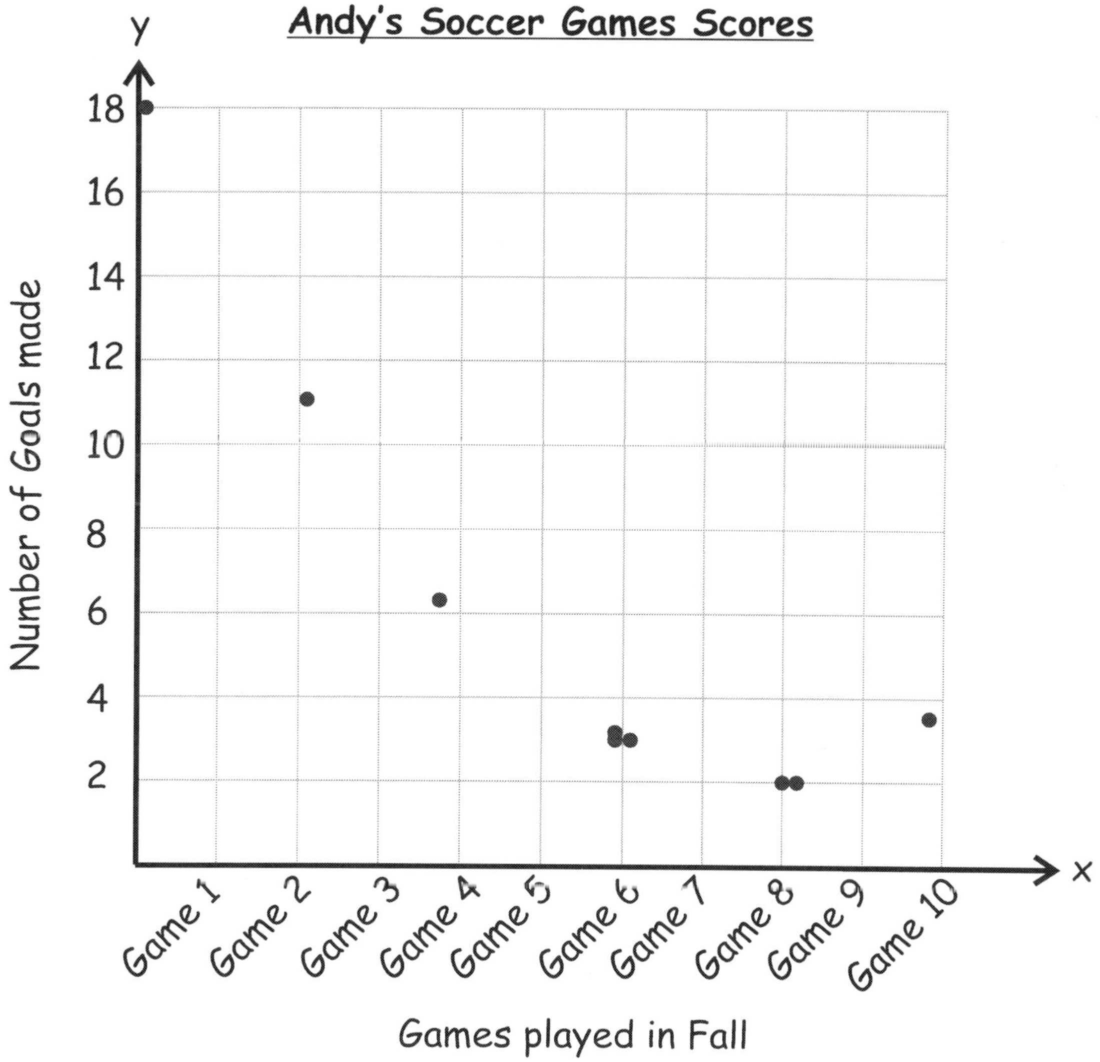

(A) Negative correlation

(B) Positive correlation

(C) No correlation

(D) None of these

SECTION D

Chuck's shoe size has increased in the last 2 years of his middle school. The number of shoes he purchased during the 2 years are plotted as below. Based on this data, answer the questions 1 - 4.

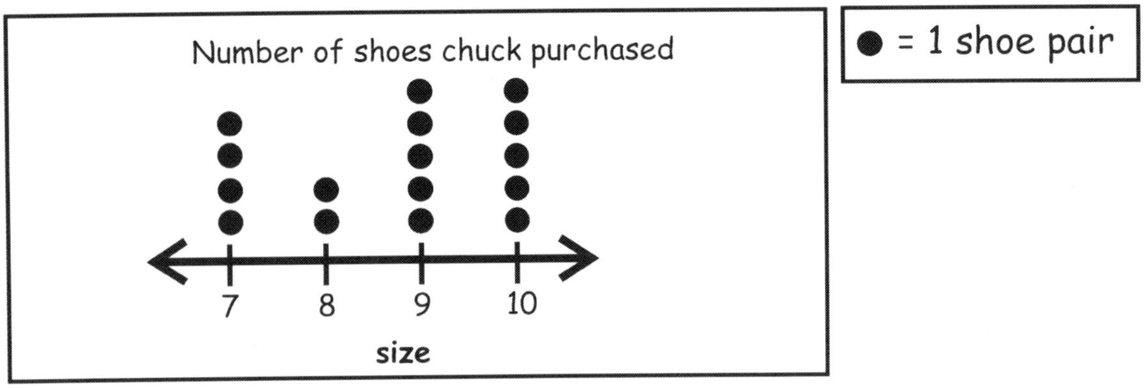

1. What is the median of the above data ?

(A) 3.5 (B) 2

(C) 3 (D) 4.5

2. What is the mean value of the above data ?

(A) 4.0 (B) 4.22

(C) 4.04 (D) 4.2

3. What is the range of the above data ?

(A) 6 (B) 8

(C) 3 (D) 4

4. What is the value of the mode from the above data ?

(A) 3 (B) 6

(C) 2 (D) 5

Dylan is analyzing car sales at his showroom across various weeks as plotted below. Based on the below data answer the questions 5 - 8

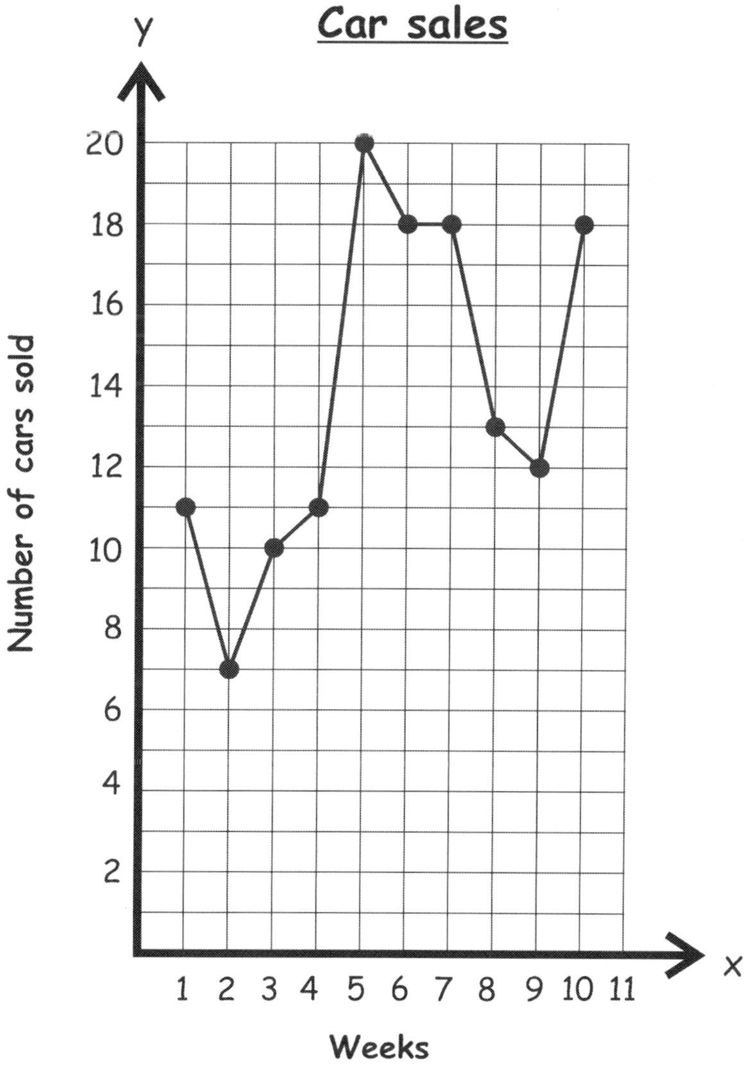

5. What is the median of the above data ?

(A) Median = 11

(B) Median = 10

(C) Median = 16

(D) Median = 12.5

6. What is the Mean of the above data ?

(A) Mean = 19.7
(B) Mean = 5.58
(C) Mean = 10.25
(D) Mean = 13.8

7. (a) What is the Range of the above data ?

(A) 12
(B) 13
(C) 16
(D) 19

(b) What is the Mode of the above data ?

(A) 18
(B) 10
(C) 15
(D) 19

8. (a) Which two weeks had the same car sales ?

(A) Weeks 2 and 5
(B) Weeks 8 and 9
(C) Weeks 6 and 7
(D) Weeks 1 and 10

(b) Which week had the highest sales ?

(A) Week 5
(B) Week 10
(C) Week 2
(D) Week 6

A number of goals made by Super Sonic Soccer Team during fall games are shown in the below graph.
Based on this data, answer the questions 9 - 12.

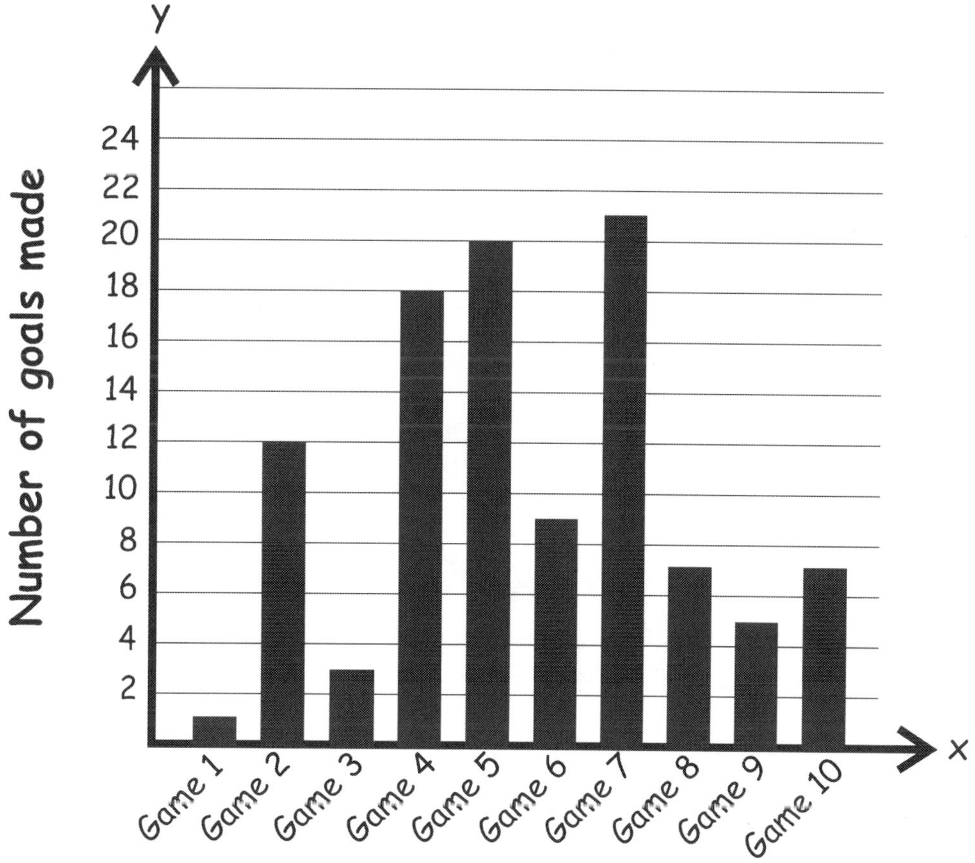

9. What is the median of the above data ?

(A) Median = 8
(B) Median = 11
(C) Median = 10.5
(D) Median = 10

10. What is the Mean of the above data ?

(A) Mean = 14.7
(B) Mean = 17.5
(C) Mean = 11.11
(D) Mean = 10.3

11. (a) What is the Range of the above data ?

(A) 18 (B) 17

(C) 20 (D) 15

(b) What is the Mode of the above data ?

(A) 9 (B) 11

(C) 7 (D) 1

12. (a) Which game had the least score ?

(A) Game 1 (B) Game 7

(C) Game 6 (D) Game 3

(b) Which two games had the same score ?

(A) Games 3 and 5 (B) Games 5 and 7

(C) Games 2 and 9 (D) Games 8 and 10

13. 15 liters of fruit punch contains 60% of grape juice. To increase the flavor nine liters of grape juice is added to it. What percentage of grape juice is in new mix ?

(A) 75% (B) 45%

(C) 70% (D) 66%

14. Find the slope of the straight line from the graph given below.

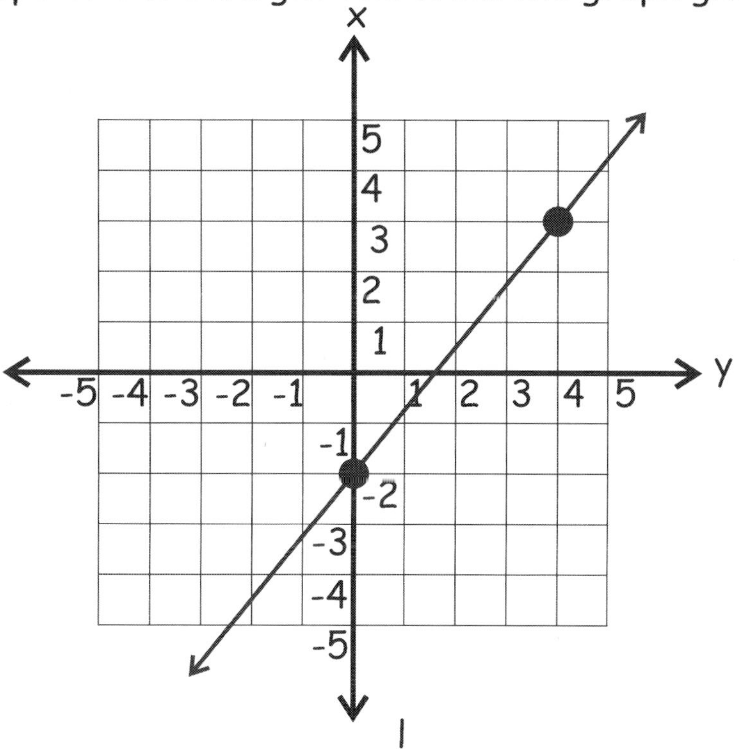

(A) $\frac{-4}{5}$ (B) $\frac{4}{5}$

(C) $\frac{5}{4}$ (D) $\frac{-5}{4}$

15. Find the missing $\angle x$ from the below figure ?

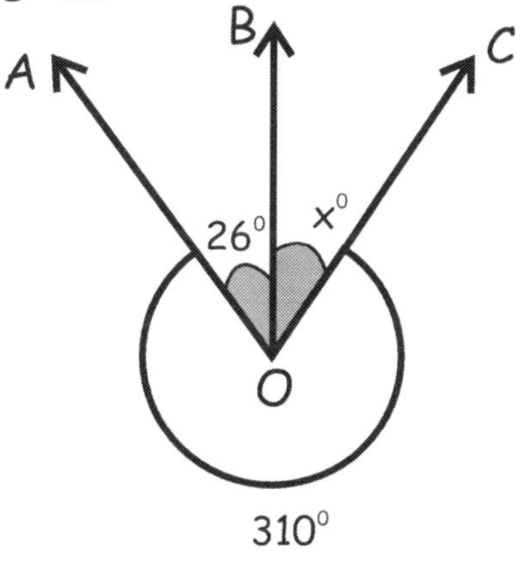

(A) 64^0 (B) 20^0

(C) 24^0 (D) 60^0

IAAT
PRACTICE TEST
TEST - 11
2
3
A
B
C

SECTION A

1. Evaluate $\frac{3}{4} \times \frac{2}{15} = \boxed{?}$

(A) $\frac{1}{10}$

(B) $\frac{1}{2}$

(C) 10

(D) 2

2. Tanya is making flower garlands with 120 flowers. She uses them in groups of 24. How many flower garlands did she made ?

(A) 5

(B) 6

(C) 10

(D) 12

3. Simplify $(-2)(-11) \times (3) + 5 \times 2 = \boxed{?}$

(A) -56

(B) 66

(C) 76

(D) -76

4. Simplify $0.004 \times 1.5 = \boxed{?}$

(A) 0.6

(B) 0.006

(C) 0.06

(D) 0.00006

5. Find the Least Common Multiple of $\boxed{60x, 45x}$

(A) 6x

(B) 15x

(C) 180x

(D) 15y

6. Which of the following sets have lowest median ?

(A) {2,5,5,2,4}

(B) {5,5,5,5,6}

(C) {9,8,7,4,10}

(D) {1,1,3,4,5}

7. Ben solved 3 puzzles at the rate of 5, 6, 7 minutes.
What is his average speed ?

(A) 5

(B) 3

(C) 12

(D) 6

8. Kasy wants to know how many times digit 5 is written from 1 through 100 ?

(A) 19

(B) 25

(C) 29

(D) 20

9. Kate wrote a number 3450. He then changed 3 to 8.
The value of the number changes by how much ?

(A) 5

(B) 5000

(C) 500

(D) 50

10. Tap Tap company is selling their watches for \$125 which were originally priced as \$150. What is the percentage of discount offered ?

(A) 17.67%

(B) 17.77%

(C) 16.07%

(D) 16.67%

11. Tim has 5 Shirts, 8 pairs of shoes and 4 ties. In how many ways can he dress for a party ?

(A) 40

(B) 160

(C) 12

(D) 32

12. Simplify the below in scientific notation

$$(6.41 \times 10^{-1})(5.62 \times 10^{2}) = \boxed{?}$$

(A) 3.602×10^{-4}

(B) 3.602×10^{2}

(C) 1.141×10^{-2}

(D) 0.1141×10^{-3}

13. Solve the below inequality

$$3p - 2p \leq 2$$

(A) $p > 2$

(B) $p \leq 2$

(C) $p \leq -14$

(D) $p \geq -24$

14. A painting is 3 in wide and 4 in tall. If it is enlarged to a height of 20 in, then how wide will it be ?

(A) 100 in

(B) 1 in

(C) 15 in

(D) 4 in

15. Express $\frac{6}{125}$ as a percentage?

(A) 6.125%

(B) 6.048%

(C) 4.8%

(D) 480%

SECTION B

1. Evaluate H, if $\frac{2}{H}=\frac{4}{3}$

(A) 3

(B) $\frac{2}{3}$

(C) 6

(D) $\frac{3}{2}$

2. A rectangle is 1 in tall and 2 in wide. If it is enlarged to a height of 7 in, then how wide is the real house ?

(A) 14 in

(B) 10 in

(C) 49 in

(D) 12 in

3. Evaluate K where $100K - 9 = 1001 - K.$

(A) 10

(B) 11

(C) 1

(D) 101

4. What is the area of a $\triangle ABC$ whose height is 10 cm and base length is 5 cm ?

(A) 5 sq cm

(B) 50 sq cm

(C) 2 sq cm

(D) 25 sq cm

5. What is the value of # in the equation $(0.8)\,.\,\# = 0.064$

(A) 8.01

(B) 0.8

(C) 0.08

(D) 0.008

6. Evaluate the expression $4X + 9Y$ when $X = 5$ and $Y = 1$.

(A) 29

(B) 27

(C) 21

(D) 26

7. If $K\#P = K + \frac{K}{P}$, then what is the value of $(3\#1)\#2$

(A) 6

(B) 9

(C) 11

(D) 19

8. Write the algebraic expression of the below:

A number cubed is 39.

(A) $3^3 = 39$

(B) $3 - n \leq 39$

(C) $\frac{3}{2}=39$

(D) $n^3 = 39$

9. Solve $-8 + 19x = 714$, and find the value of x.

(A) 38

(B) 21

(C) 19

(D) -22

10. Jessica was going to sell all of her stamp collection to buy a video game. After selling half of them she changed her mind. She then bought eighteen more. How many did she start with if she now has 33 ?

(A) 48

(B) 66

(C) 84

(D) 30

11. Solve the below proportion and find the value of 'a'.

$$\frac{17}{12} = \frac{a}{19}$$

(A) $\frac{323}{12}$

(B) -2

(C) $\frac{-4}{3}$

(D) 11

12. Express the below algebraic expression in words:

$$2^n > 20$$

(A) The n power of 2 is greater than 20

(B) The 2nd power of n is greater than 20

(C) The product of n and 2 is greater than 20

(D) n less than 2 is greater than 20

13. Evaluate p - (p - q) where p = 6, and q = 1

(A) 6

(B) 1

(C) 7

(D) 3

14. Simplify $(5x^2 - 3x^4) - (2x^2 - 3 - 3x^4 - x^3)$

(A) $x^3 - 2x^2 - 4$

(B) $x^3 - 2x^2 + 3$

(C) $x^3 - 2x^2 - 1$

(D) $x^3 + 3x^2 + 3$

15. A model house has a scale of 4.5 cm : 18 m. If the model house is 4 cm wide, then how wide is the real house ?

(A) 22 m

(B) 1.1 m

(C) 16 m

(D) 9 m

SECTION C

1. Togethe Sarah and Juhan ate twenty six crackers. If Juan ate four more than Sarah, how many crackers did Sarah ate ?

(A) $26 = 2y + 4; y=11$ (B) $26 = y+4; y=22$

(C) $26 = 2y-4; y=15$ (D) $26 = 4y+2; y=6$

2. Evaluate $5(-2 + 7x) = 270$ for 'x'

(A) -5 (B) -13

(C) 8 (D) 2

3. Olivia has a desert stand at a school fund raiser. For each brownie she sells two cupcakes. If she has sold 76 cupcakes then how many brownie did she sold ?

(A) 38 (B) 33

(C) 29 (D) 41

4. Evaluate $\frac{12}{7} \div \frac{3}{2}$

(A) $\frac{1}{7}$ (B) $\frac{8}{7}$

(C) $\frac{5}{7}$ (D) $\frac{2}{7}$

5. Evaluate 12.7 ÷ -12.7

(A) -9

(B) -1

(C) 2

(D) 5

6. Which of the below is not a function ?

(A) {(-8, 5), (0, 1), (7, 6), (12, 3), (23, 34)}

(B) {(2, 3), (4, 15), (5, 24), (6, 35), (7, 48)}

(C) {(9, 8), ((2, 3), (6, 7), (2, 7), (8, 13)}

(D) {(3, 9), (4, 16), (5, 25), (6, 36), (7, 35)}

7. A ladder that is 20 ft tall casts a shadow that is 8 ft long. Find the length of the shadow that a 5 ft woman casts.

(A) 2 ft

(B) 10 ft

(C) 16 ft

(D) 5 ft

8. Write the algebraic expression for the statement:

The quotient of v and 3 is 14.

(A) 3 + v = 14

(B) $\frac{3}{v}=14$

(C) 3v = 14

(D) $\frac{v}{3}=14$

9. Tim rolls a 11 sided dice. What is the probability of rolling a number higher than 7 ?

(A) $\frac{7}{11}$

(B) $\frac{5}{11}$

(C) $\frac{6}{11}$

(D) $\frac{4}{11}$

10. Ishaan can read 100 pages in 3 hour. He reads P pages in 7 hours. Using proportion make an equation.

(A) $\frac{3}{100}=\frac{7}{P}$

(B) $\frac{3}{100}=\frac{P}{7}$

(C) $\frac{100}{3}=\frac{7}{P}$

(D) $21 = 100\,P$

11. Write the algebraic expression of the statement $n + 7 = 49.$

(A) A number cubed is equal to 49

(B) Twice 7 is equal to 49

(C) Twice a number is equal to 49

(D) A number increased by 7 is equal to 49

12. Evaluate $H^2 + H$ where $H^{-3} = -\frac{1}{8}$

(A) $\frac{1}{4}$

(B) 2

(C) -2

(D) 4

13. Rik had $24 to buy four notebooks. After buying them he had $8 left. What is the price of each note book?

(A) $4 (B) $2

(C) $1 (D) $8

14. Find the missing value from the below table.

INPUT	OUTPUT
11	6
5	0
7	2
9	?
10	5

(A) 0 (B) 8

(C) 4 (D) 3

15. Find the correlation in the data given below.

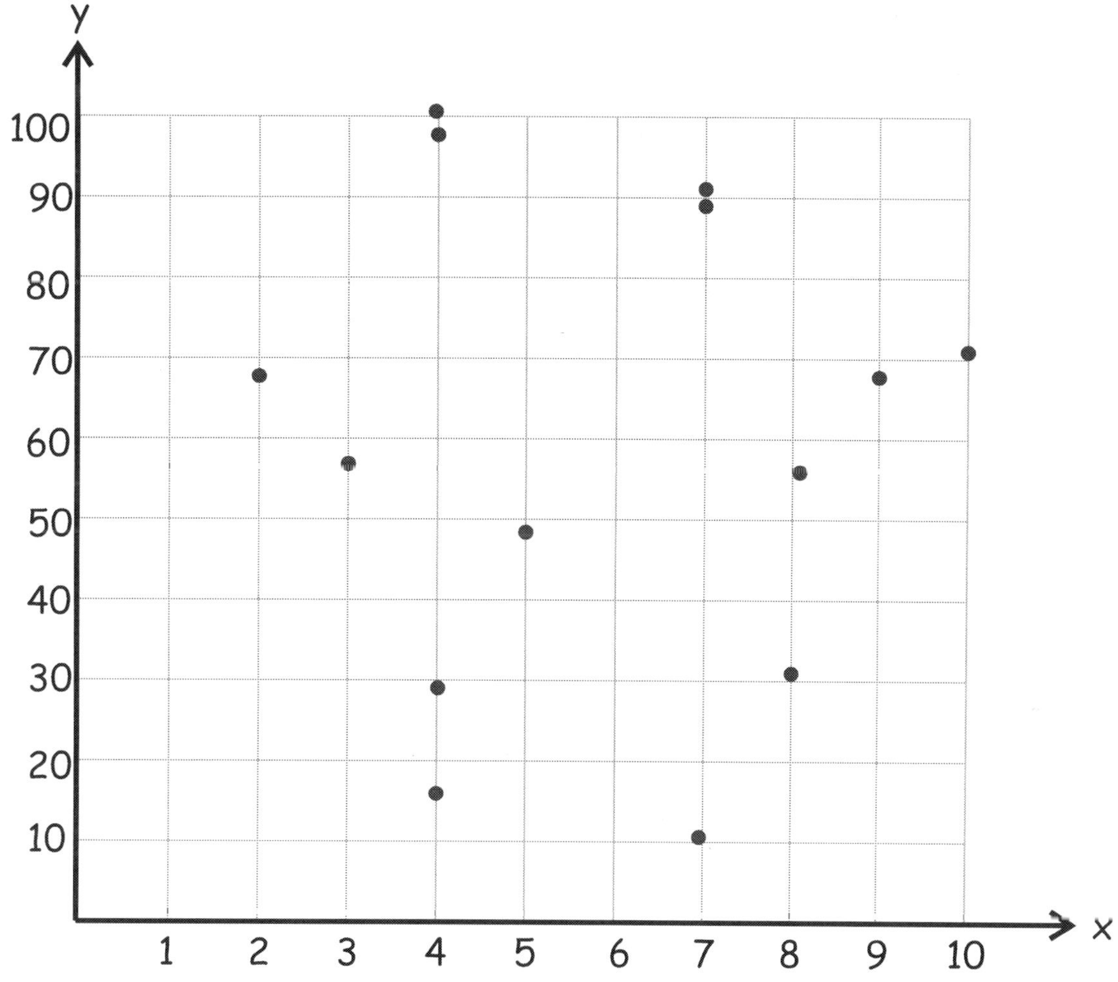

(A) Negative correlation

(B) No correlation

(C) Positive correlation

(D) None of these

SECTION D

Based on the graph given below, Answers the questions 1 - 4

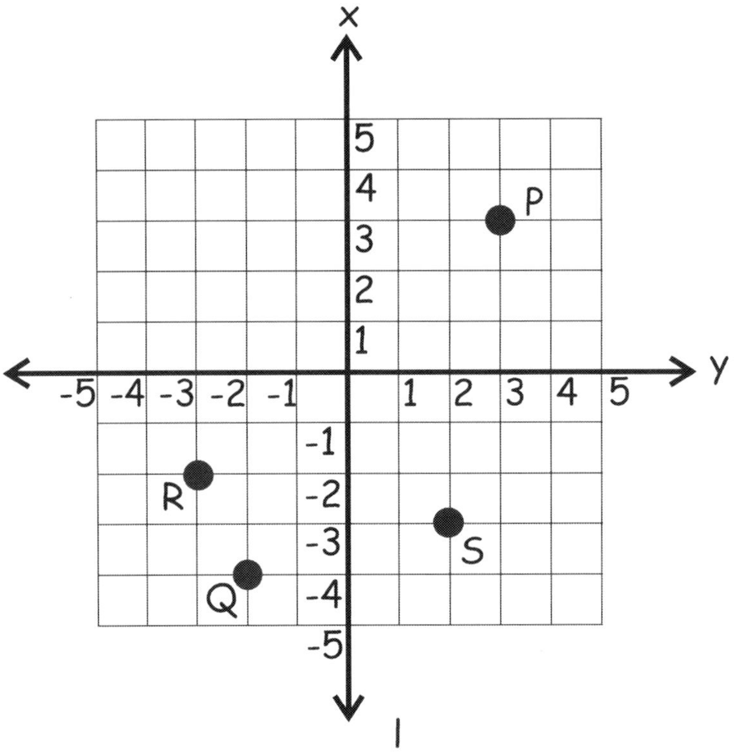

1 (a) The point Q lies in which quadrant ?

(A) Quadrant I (B) Quadrant II

(C) Quadrant III (D) Quadrant IV

(b) The point S lies in which quadrant ?

(A) Quadrant I (B) Quadrant II

(C) Quadrant III (D) Quadrant IV

2. Which of the below represents the coordinates of point R ?

(A) (-3,-2) (B) (3,-2)

(C) (2,-3) (D) (2,3)

3. Which of the below represents the coordinates of P ?

(A) (3,3) (B) (3,-3)

(C) (-3,-3) (D) (-3,3)

4. What is the additive inverse of the below ?

5067

(A) 0 (B) -5067

(C) $\frac{-1}{5067}$ (D) $\frac{1}{5067}$

5. Solve ; $115 = 5(3n + 8)$

(A) 6 (B) -9

(C) 2 (D) 5

6. Several people are standing in a straight line to buy concert tickets. Starting at one end of the line, Bill is counted as the 15th person, and starting at the other end, he is counted as the 37th person. How many people are in the line ?

(A) 48 (B) 22

(C) 45 (D) 51

7. Lola is getting ready for a music show. Everyday she practices singing and piano in the ratio of 6 to 4. Today she practices nine hours of singing, then how many hours of piano does she have practiced ?

(A) 5 (B) 4

(C) 7 (D) 6

8. A long distance call costs x cents for the first minute and y cents for each additional minute, how much would it cost for a five minute call ?

(A) 5xy (B) x + 5y

(C) $\frac{xy}{5}$ (D) x + 4y

9. Sally gets paid x dollars per hour for a 40-hour work week and y dollars for each hour she works over 40 hours. How much did Sally earns if she worked 48 hours ?

(A) 48xy (B) 40y + 8x

(C) 40x + 8y (D) x + 4y

10. Which of the below is the rational number ?

(A) $\sqrt{36}$ (B) $\sqrt{35}$

(C) $\sqrt{105}$ (D) $\sqrt{522}$

11. Find the value of $\angle POQ$ from the figure given below

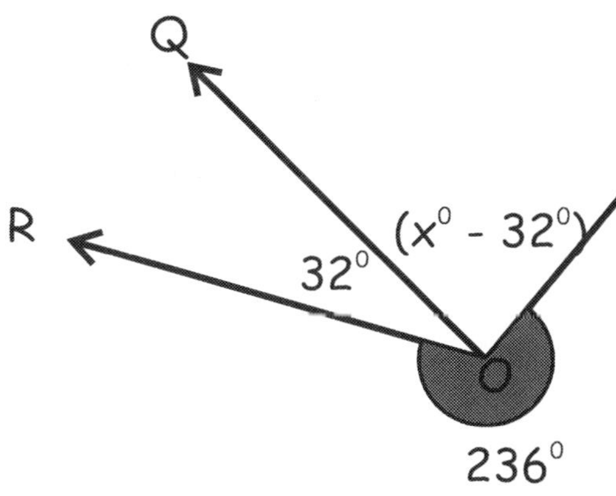

(A) 98°	(B) 118°

(C) 102°	(D) 60°

12. Find the value of X from the figure given below

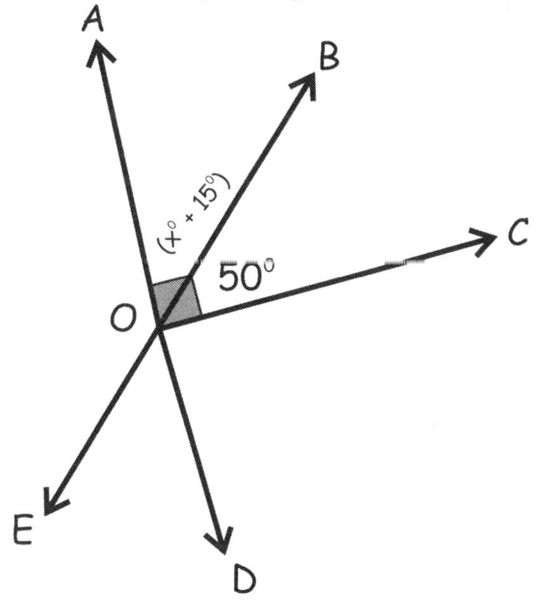

(A) 25°	(B) 15°

(C) 20°	(D) 18°

13. If ab = 28 and if the value of 'a' increases, what will happen to the value of 'b' ?

(A) Increases
(B) Remains same
(C) Decreases
(D) None

14. Given 2a - b = 8. If the value of a decreases, what will happen to the value of b ?

(A) Increases
(B) Decreases
(C) Remains same
(D) None

15. Evaluate

$$-2|x - 12| = -18$$

(A) {6,18}
(B) {-4,6}
(C) $\{\frac{2}{3}, 6\}$
(D) {2,22}

IAAT
PRACTICE TEST
TEST - 12

SECTION A

1. Evaluate $\frac{5}{11} \times \frac{33}{100} = \boxed{?}$

(A) $\frac{1}{20}$

(B) $\frac{3}{10}$

(C) $\frac{1}{10}$

(D) $\frac{3}{20}$

2. Tony is planting rose shrubs in his square shaped rose garden. If he plants 4 of them in a row and the distance between any two plants is 5 ft, then what is the area of his garden ?

(A) 400 sq ft

(B) 40 sq ft

(C) 225 sq ft

(D) 15 ft

3. Simplify $(-11)\,(-3) \times (2) + (-9) \times (5) \;\; = \boxed{?}$

(A) 21

(B) 19

(C) 24

(D) 111

4. Simplify $0.240 \times 3.2 \; = \boxed{?}$

(A) 7.238

(B) 7.68

(C) 0.768

(D) 0.748

5. Find the Least Common Multiple of $22y^2, 77y^2$

(A) $9y^2$
(B) $11y^2$
(C) $154y^2$
(D) $7y^2$

6. Which of the following sets have greatest average ?

(A) {2,5,5,2,4}
(B) {5,5,5,5,6}
(C) {9,8,7,4,10}
(D) {1,1,3,4,5}

7. Beth is driving car at the rate of 25 miles/hr for the first hour and then 40 miles per hour for next hour. What is his approximate average speed ?

(A) 31 m/hr
(B) 31 m/hr
(C) 33 m/hr
(D) 35 m/hr

8. Kasy wants to know how many times digit 2 is written from 1 through 50. Can you help?

(A) 14
(B) 35
(C) 20
(D) 19

9. Andy wrote a number 74,321. He then changed 7 to 5. By how much does the value of the number changes ?

(A) 20,000 (B) 5,000

(C) 7,000 (D) 3,000

10. Creative Art company is selling their art pieces for $75 which were originally priced as $125. What is the percentage of discount offered ?

(A) 60% (B) 50%

(C) 25% (D) 40%

11. Dan has 9 pants, 7 shirts and 5 watches. In how many ways can he dress ?

(A) 305 ways (B) 311 ways

(C) 315 ways (D) 325 ways

12. Simplify the below in scientific notation

$$(1.5 \times 10^{-2})(205 \times 10^{-5}) = \boxed{?}$$

(A) 37.5×10^{-7} (B) 3.75×10^{-7}

(C) 0.0375×10^{-7} (D) 0.375×10^{-4}

13. Solve the below inequality.

$$-6r - 3r \leq -18$$

(A) $r \leq -15$

(B) $r \leq -14$

(C) $r \geq 2$

(D) $r \leq 2$

14. If a*b = 3(a + b), what is -4*9

(A) 15

(B) 11

(C) -10

(D) 9

15. Express $\frac{17}{25}$ as a percentage.

(A) 158%

(B) 580%

(C) 0.68%

(D) 68%

SECTION B

1. Evaluate H, if $\frac{12}{H} = \frac{6}{15}$

(A) 30

(B) 21

(C) 18

(D) 27

2. Evaluate $3(1 - 4x) = 99$

(A) 10

(B) 8

(C) 9

(D) -8

3. Evaluate K where $99K - 9 = 1 - K.$

(A) $\frac{1}{5}$

(B) $\frac{1}{10}$

(C) $\frac{2}{5}$

(D) $\frac{3}{7}$

4. Volume of a cube is 125 cubic inches. What is the length of the side ?

(A) 15

(B) 5

(C) 25

(D) 1

5. What is the value of # in the equation $(0.13) \cdot \# = 1.69$

(A) 13

(B) 0.13

(C) 1.3

(D) 23

6. Evaluate the expression $7X + 5Y$ when $X = 5$ and $Y = 7$.

(A) 35 (B) 12

(C) 0 (D) 70

7. If $K\#P = K \times K \times \frac{2}{P}$, then what is the value of $(3\#1)\#3$

(A) 206 (B) 216

(C) 316 (D) 226

8. Write the algebraic expression of the below:

The product of a number and 7 is greater than or equal to 46

(A) $n - 7 < 46$ (B) $n \cdot 7 \geq 46$

(C) $7 - n \geq 46$ (D) $n - 7$

9. Solve $\frac{-16+p}{-5}=2$, find the value of p ?

(A) -23 (B) 6

(C) 3 (D) 38

10. Shawna spent half of her weekly allowance on clothes. To earn more money, her parents let her wash the car for \$10. What is her weekly allowance if she ended with \$17 ?

(A) \$3.50 (B) \$24

(C) \$12 (D) \$14

11. Solve the below proportion and find the value of x.

$$\frac{8}{16} = -\frac{5}{x}$$

(A) $-\frac{1}{2}$ (B) $1\frac{1}{2}$

(C) -2 (D) -10

12. Express the below algebraic expression in words:

$$2n \geq 37$$

(A) 2 more than n is greater than or equal to 37

(B) The difference of n and 2 is greater than or equal to 37

(C) Twice n is greater than or equal to 37

(D) 2 less than n is greater than or equal to 37

13. Evaluate 2ab where a = 3, and b = 5

(A) 27

(B) 30

(C) 34

(D) 25

14. Simplify $(3x^3 - 1) - (5 - 2x^3 + x^2 + x^4)$

(A) $-x^4 + 5x^3 - x^2 - 6$

(B) $-2x^4 + 8x^3 - 6x^2 - 6$

(C) $-x^4 + 5x^3 - 6x^2 - 6$

(D) $-2x^4 + 5x^3 - 6x^2 - 6$

15. Simplify $\frac{-6.5}{-0.4}$

(A) 11.45

(B) 14

(C) 16.25

(D) 15

SECTION C

1. Evaluate $\frac{1}{2} \div \frac{1}{2}$

(A) 5 (B) 12

(C) 3 (D) 1

2. Evaluate $\boxed{6(1 + 8x) = 150}$ for 'x'.

(A) -2 (B) 3

(C) 5 (D) 7

3. Yummy ice cream stores sells 3 vanilla ice creams for every 2 chocolate ice creams. Today if they have sold 96 chocolate ice creams then how many vanilla ice creams were sold ?

(A) 144 (B) 150

(C) 145 (D) 165

4. Two less than the product of a number and -5 is -42. Write the expression and find the number.

(A) $\frac{x}{-5} - 2 = -42; x=200$ (B) $x - 2 = -42; x=-40$

(C) $-5x-2 = -42; x=8$ (D) $-5x - 2 = -42; x=\frac{44}{5}$

5. A triangle is 2 in tall and 1 in wide. If it is enlarged to a height of 20 in, then how wide will it be ?

(A) 100 in

(B) 2 in

(C) 11 in

(D) 10 in

6. Which of the below is not a function?

(A) {(-3, 3), (-2, 5), (-3, 7), (9, 0), (5, 0)}

(B) {(-4, 0), (1, 5), (-5, 9), (-6, -11), (7, 11)}

(C) {(-6, 39), (3, 10), (4, 17), (5, 26), (6, 39)}

(D) {(-9, 0), (0, 8), (5, 9), (6, 10), (7, 9)}

7. A tent that is 10 ft tall casts a shadow that is 5 ft long. Find the height of a petrified stump that casts a 4 ft shadow.

(A) 5 ft

(B) 8 ft

(C) 20 ft

(D) 50 ft

8. Write the algebraic expression of the statement:

Twice z is less than 16

(A) $z - 2 \leq 16$

(B) $2 - z \geq 16$

(C) $2z < 16$

(D) $2^z < 16$

9. Cathy rolls a 14 sided dice. What is the probability of rolling a number less than 7 ?

(A) $\frac{5}{14}$ (B) $\frac{1}{7}$

(C) $\frac{4}{7}$ (D) $\frac{3}{7}$

10. Tej can read 10 stories in 2 and half hours . He reads S stories in 24 hours. Using proportion make an equation.

(A) $\frac{S}{24} = \frac{1}{4}$ (B) $\frac{25}{10} = \frac{S}{24}$

(C) $4 = \frac{24}{S}$ (D) $\frac{2}{10} = \frac{24}{S}$

11. Write the algebraic expression of $r \cdot 6 = 21$

(A) r times 6 is equal to 21

(B) r increased by 6 is equal to 21

(C) r more than 6 is equal to 21

(D) 6 to the r is equal to 21

12. Evaluate $7H^2 + 4H + 12$ where $H^{10001} = -1$

(A) 15 (B) 23

(C) 19 (D) 1

13. Cathy's Grand pa told her, if 300 reduced by 3 times his age is 138. How old is Cathy's Grand Pa?

(A) 38

(B) 69

(C) 51

(D) 54

14. Find the missing value from the below table.

INPUT	OUTPUT
-5	-3
-3	-1
0	2
4	?
7	9

(A) 4

(B) 5

(C) 6

(D) -7

15. Find the correlation in the data given below.

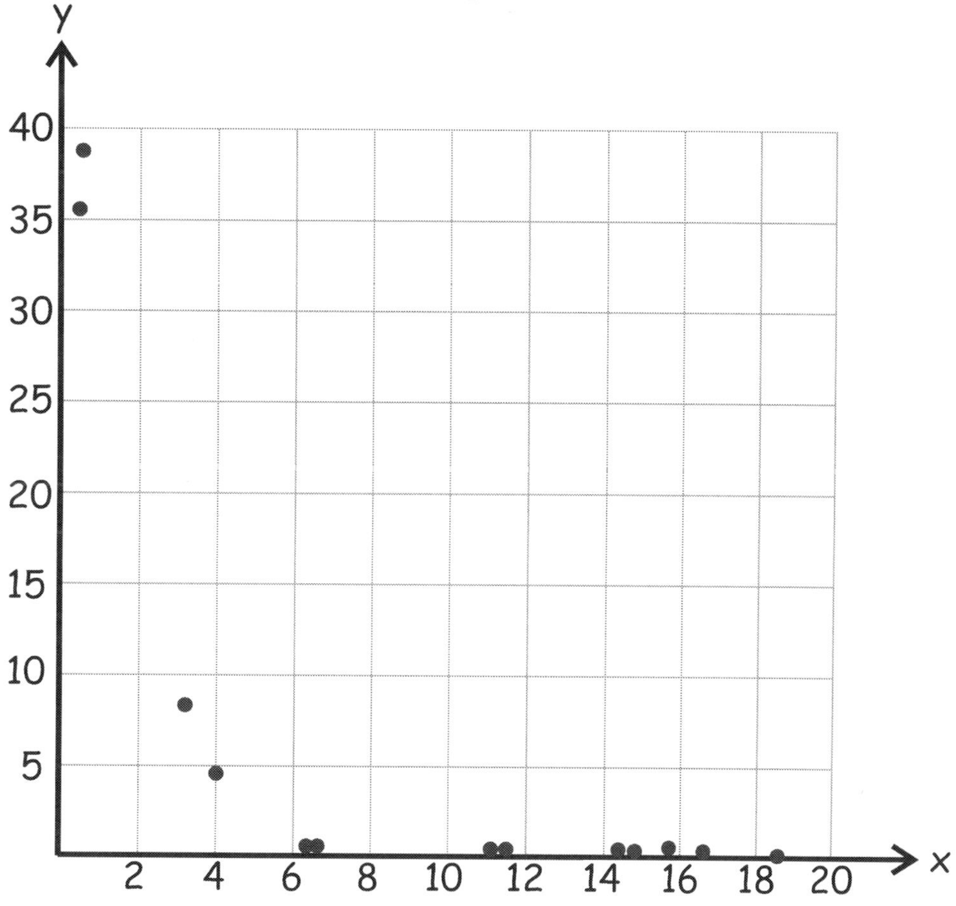

(A) Negative correlation

(B) No correlation

(C) Positive correlation

(D) None of these

SECTION D

Based on the graph given below, Answers the questions 1 - 4

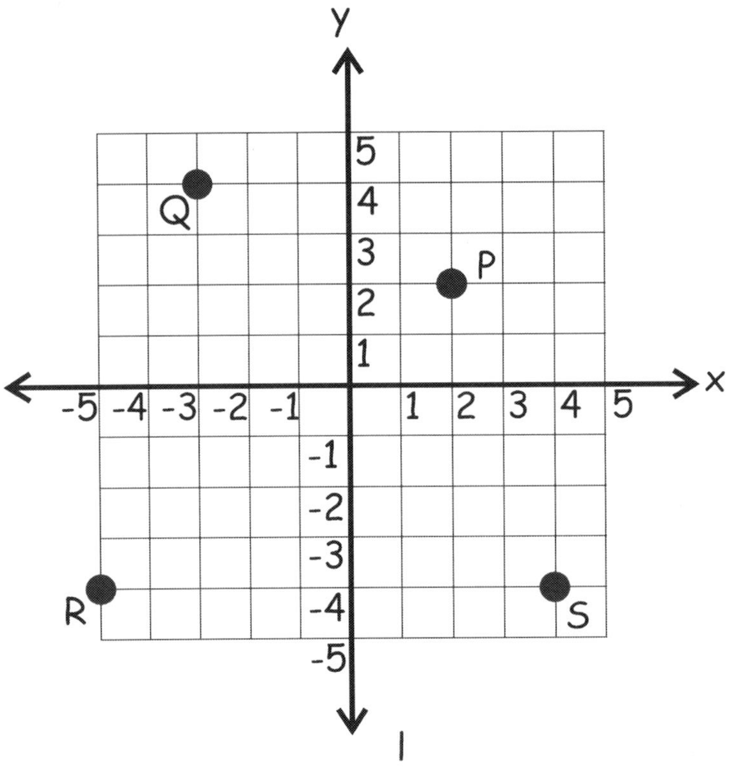

1 (a) The point Q lies in which quadrant ?

(A) Quadrant I (B) Quadrant II

(C) Quadrant III (D) Quadrant IV

(b) The point P lies in which quadrant ?

(A) Quadrant I (B) Quadrant II

(C) Quadrant III (D) Quadrant IV

2. Which of the below represents the coordinates of point R ?

(A) (-5,4)
(B) (5,-2)
(C) (-4,-3)
(D) (-5,-4)

3. Which of the below represents the coordinates of P ?

(A) (2,5)
(B) (-2,-2)
(C) (2,2)
(D) (-5,0)

4. The difference between four times a number and six is -46. Write the expression and find the number ?

(A) $\frac{x}{4} - 6 = -46;\ x = -160$
(B) $4x - 6 = -46;\ x = -10$
(C) $4x + 6 = -46;\ x = -13$
(D) $4x - 6 = -46;\ x = -13$

5. A rectangle is 4 inches tall and two inches wide. If it is enlarged to a width of 4 inches, then how tall would it be ?

(A) 11 inches
(B) 8 inches
(C) 16 inches
(D) 12 inches

6. Cory purchased a frying pan that was on sale for 30% off. She saved \$3.75 with the sale. What was the original price of the frying pan ?

(A) \$10.90 (B) \$9.25

(C) \$11.25 (D) \$12.50

7. The sum of two consecutive odd integers is -112. What is the larger integer ?

(A) -57 (B) -48

(C) 55 (D) 57

8. John has a lemonade stand at a local fair. For every four small glasses he sells ten big glasses. If he has sold 8 small glasses in last hour then how many big glasses of lemonade did he sold ?

(A) 26 (B) 16

(C) 34 (D) 20

9. A painting is 3 in tall and 4 in wide. If it is enlarged to a height of 15 in, then how wide will it be ?

(A) 3 in (B) 75 in

(C) 1 in (D) 20 in

10. Which of the below is the rational number ?

(A) $\sqrt{136}$ (B) $\sqrt{121}$

(C) $\sqrt{1025}$ (D) $\sqrt{47}$

11. Find the value of $\angle POQ$ from the figure given below

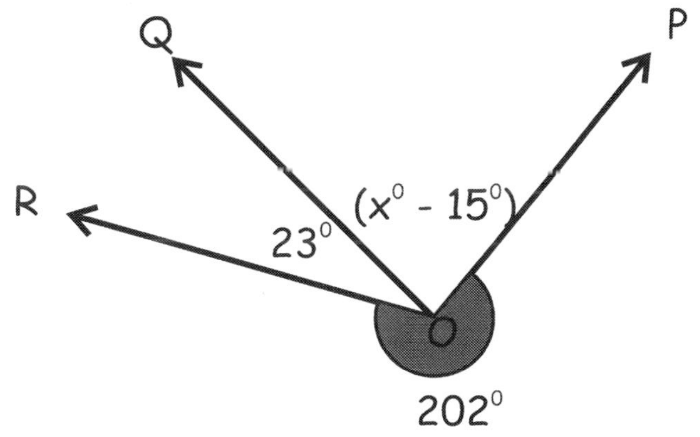

(A) 130° (B) 135°

(C) 120° (C) 102°

12. Evaluate X from the below

$$|X - 8| = 4$$

(A) {12,4} (B) {-12,6}

(C) {5,-9} (D) {3,4}

13. If 5a-3b = 44 and if the value of 'a' increases, what will happen to the value of 'b' ?

(A) Decreases (B) Remains same

(C) Increases (D) None

14. If $\frac{2a + 1}{b} = 4$ and if the value of 'a' increases, what will happen to the value of 'b' ?

(A) Decreases (B) Remains same

(C) Increases (D) None

15. Find the value of X

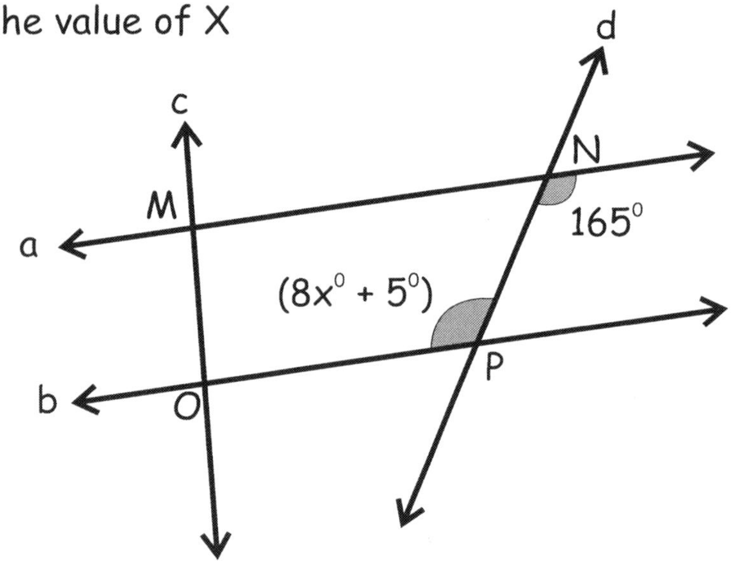

(A) 20^0 (B) 15^0

(C) 35^0 (D) 25^0

IAAT
PRACTICE TEST
TEST - 13

SECTION A

1. Find the percentage change from 39 to 29 to the nearest tenth.

(A) 134.5% decrease (B) 34.5% decrease

(C) 10% decrease (D) 25.6% decrease

2. The original price of a scarf is $ 8.99 and the store offers a discount of 36%. What is the sale price of the scarf ?

(A) $ 12.23 (B) $ 8.54

(C) $ 5.57 (D) $ 3.24

3. Find the value of H satisfying the below equation

$$\boxed{5H-3T=11} \quad \text{where } T=1$$

(A) $\frac{14}{5}$ (B) $\frac{19}{5}$

(C) $\frac{17}{4}$ (D) $\frac{11}{7}$

4. Find the value of H satisfying the below equation

$$\boxed{3T-H=1} \quad \text{where } T=1$$

(A) 2 (B) -10

(C) 5 (D) -7

5. Find the value of H satisfying the below equation

$$-3 < 2H < 8$$

(A) 2 (B) 10

(C) 22 (D) -8

6. Find the value of H satisfying the below equation

$$-2 < H + 1 < 9$$

(A) 1 (B) 4

(C) 10 (D) 77

7. Evaluate : $329 = -7(1 + 8a)$

(A) -13 (B) -9

(C) -6 (D) 4

8. Evaluate the below

$$-5 + 2x = -43$$

(A) -19 (B) 19

(C) 12 (D) -8

9. Simplify the below

$$(2p^2 - 4p) + (p + 7p^4) + (p^2 - 4p)$$

(A) $7p^4 - 3p^2 + 7p$

(B) $7p^4 + 2p^2 - 9p$

(C) $7p^4 - 2p^2 + 7p$

(D) $7p^4 + 3p^2 - 7p$

10. Evaluate

$$\frac{A}{15} + 7 = 6$$

(A) -15

(B) 15

(C) -225

(D) 225

11. Evaluate : $2.4 \div 8$

(A) 0.5

(B) 0.3

(C) 0.03

(D) 3

12. Evaluate the below

$$x - y + y^2 - x$$ Where $x = 11$, $y = 5$

(A) 36

(B) 25

(C) 5

(D) 20

13. Evaluate : $\frac{1}{5} + 2\frac{3}{4}$

(A) $2\frac{19}{20}$ (B) $2\frac{7}{60}$

(C) $\frac{13}{40}$ (C) $7\frac{37}{60}$

14. Write the verbal expression for the below

$n + 7 < 13$

(A) the seventh power of n is less than 13

(B) n increased by 7 is less than 13

(C) twice n is less than 13

(D) twice 7 is less than 13

15. The sum of three cosecutive number is 36. What is the largest of these numbers ?

(A) 7 (B) 11

(C) 15 (D) 13

SECTION B

1. Which of the following expressions is equivalent to

$$a^8$$

(A) $a \times a \times a \times a \times a \times a \times a$

(B) $a \times a \times a \times a \times a \times a \times a$

(C) $a \times a \times a \times a \times a \times a$

(D) $a \times a \times a \times a \times a$

Based on the data given below, answer the question 2-4

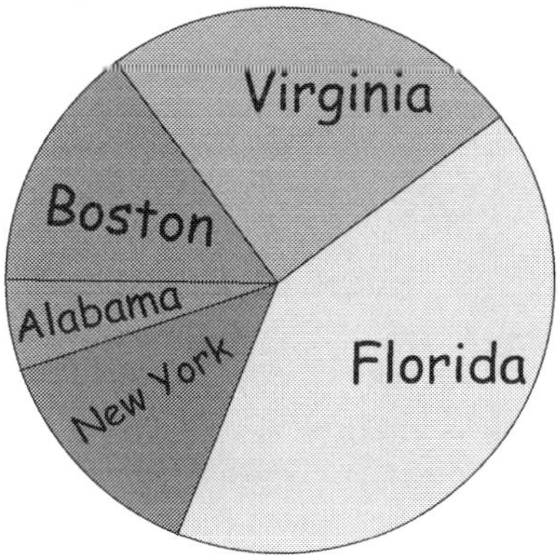

2. Which state is most voted ?

(A) New york

(B) Boston

(C) Florida

(D) Alabama

3. Which state is least voted ?

(A) New Jersey

(B) Alabama

(C) New York

(D) Florida

4. Which two states have equal votes?

(A) Boston and New York
(B) Boston and Virginia
(C) New York and Alabama
(D) Florida and Alabama

5. Which of the below point lies in fourth quadrant ?

(A) (-7,-11)
(B) (7,11)
(C) (7,-11)
(D) (-7,11)

6. Which of the below point lies in third quadrant ?

(A) (-4,12)
(B) (-4,-12)
(C) (4,12)
(D) (4,-12)

7. Simplify the below

$$(r \times r \times r \times r \times r \times r)(r \times r \times r \times r \times r)$$

(A) r^{10}
(B) r^{13}
(C) r^{8}
(D) r^{11}

8. The ratio of pencils to pen in a box is 8:5. If there are 16 pencils then, how many pencils are there ?

(A) 10
(B) 16
(C) 7
(D) 21

9. Which of the following has the smallest value ?

(A) $|-237|$

(B) $|-11|$

(C) $-|93|$

(D) $-|101|$

10. Seven times a number plus the sum of three and twice a number is 30. Write the expression and find the number.

(A) 7n+2n+3=30;n=3

(B) 7n+2n-3=30;n= $\frac{11}{3}$

(C) 7n-2n-3=30;n= $\frac{27}{5}$

(D) 7+2n+3=30;n=10

11. Evaluate for p

$$7 \times 236 = (7 \times 200) + (7 \times 30) + (7 \times p)$$

(A) 2

(B) 5

(C) 6

(D) 1

12. The below equation is an example of

x = y ; y = z then x = z

(A) Distributive property

(B) Reflexive property

(C) Associative property

(D) Symmetric property

13. A model rocket has a scale of 4.5 cm : 14.9m. If the model rocket is 4 cm tall, then how tall is the real rocket ?

(A) 1.2 m

(B) 18.9 m

(C) 13.2 m

(D) 0.9 m

14. How many feet are in five yards ?

(A) 5

(B) 15

(C) 3

(D) 20

15. Which of the below equations is represented by the figure give below ?

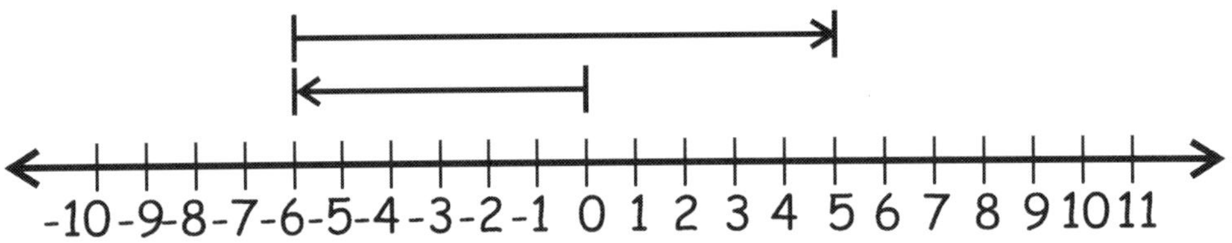

(A) -11 + 0

(B) -6 + 11

(C) -6 + 6

(D) -11 + 6

SECTION C

1. Peter purchased 14 new baseball cards for his collection. This increased the size of his collection by 35%. How many baseball cards does Peter now have ?

(A) 40 (B) 54

(C) 27 (D) 34

2. Melissa is four times as old as Jim. Pat is 5 years older than Melissa. If Jim is y years old, how old is Pat ?

(A) $5y + 4$ (B) $4 \times 5y$

(C) $y + 5$ (D) $4y + 5$

3. Ella is combining two 6-inch pieces of wood with a piece that measures 4 inches. How many total inches of wood does he have ?

(A) 10 inches (B) 8 inches

(C) 16 inches (D) 12 inches

4. $\triangle ABC$ is translated to 4 units down. Find the coordinates of each vertex after the transformation.

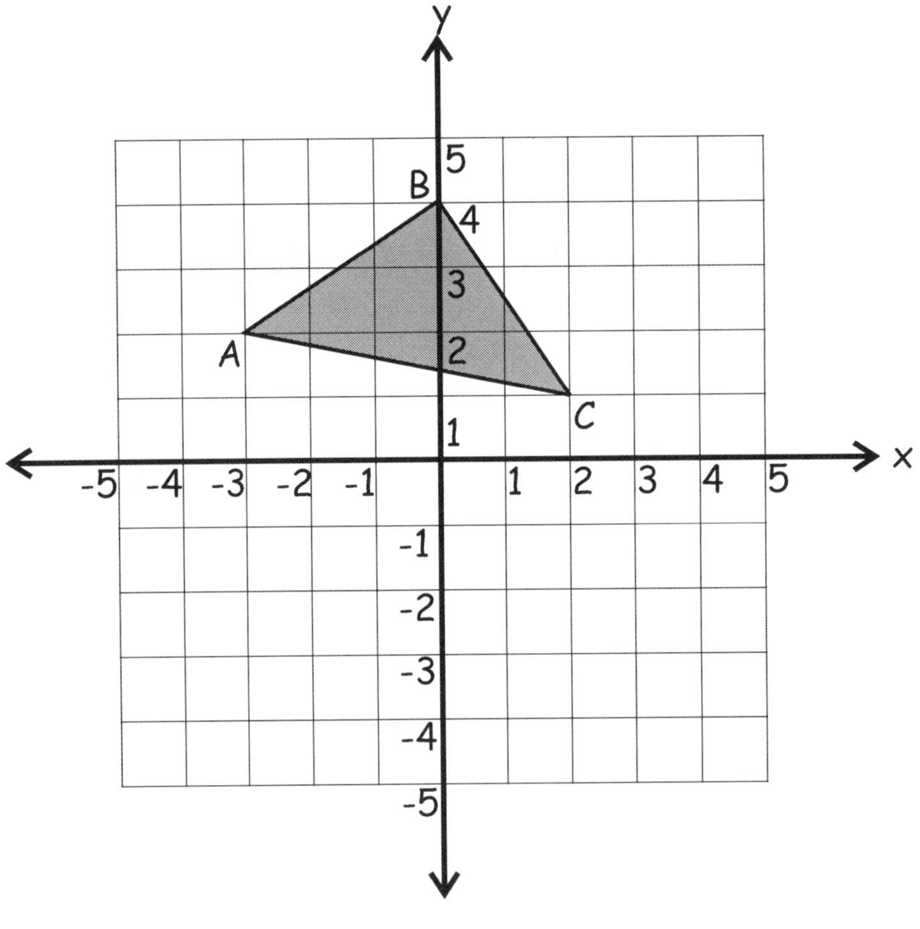

(A) A'(0,-2),B'(-3,-4),C'(-5,1)

(B) A'(-4,2),B'(-1,4),C'(1,1)

(C) A'(-5,-2),B'(-2,-4),C'(0,-1)

(D) A'(-3,-2),B'(0,0),C'(2,-3)

5. Rotate the $\triangle ABC$ 180° about origin. Find the coordinates of each vertex after the transformation.

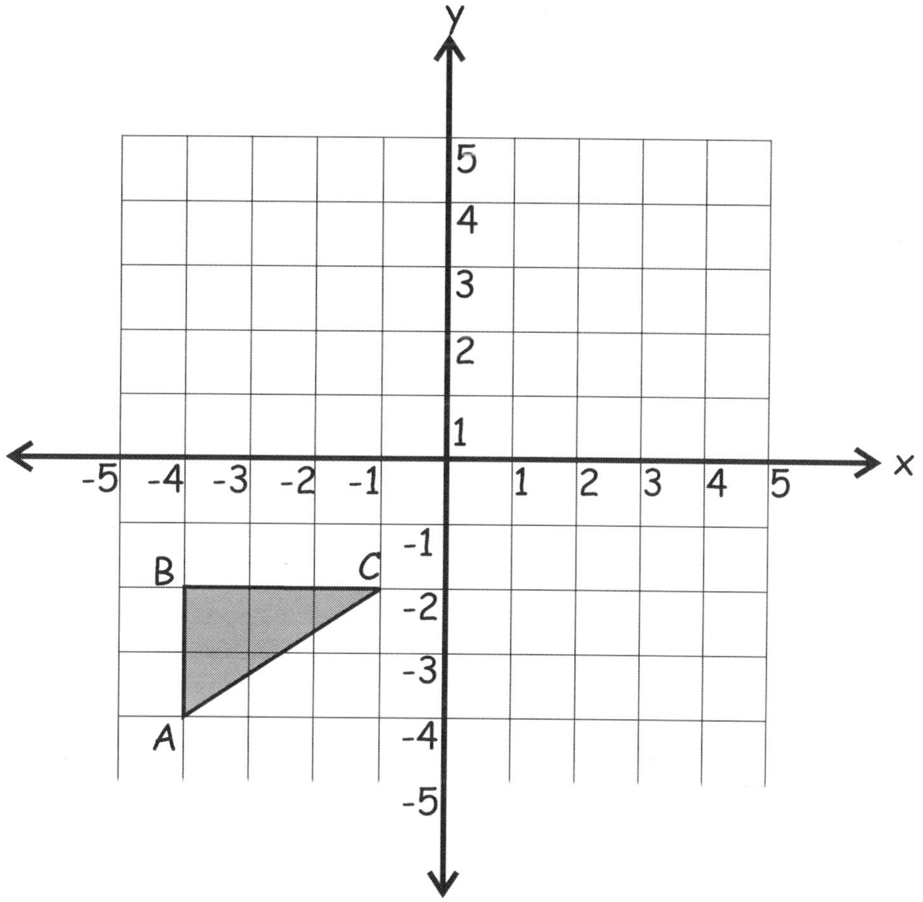

(A) A'(4,4),B'(-2,4),C'(-2,-1)

(B) A'(4,4),B'(4,2),C'(1,2)

(C) A'(4,-4),B'(2,-4),C'(2,-1)

(D) A'(4,2),B'(1,-2),C'(4,0)

Legacy elementary school students voted on various sports available at school as shown in the below graph. Answer the questions 6-10

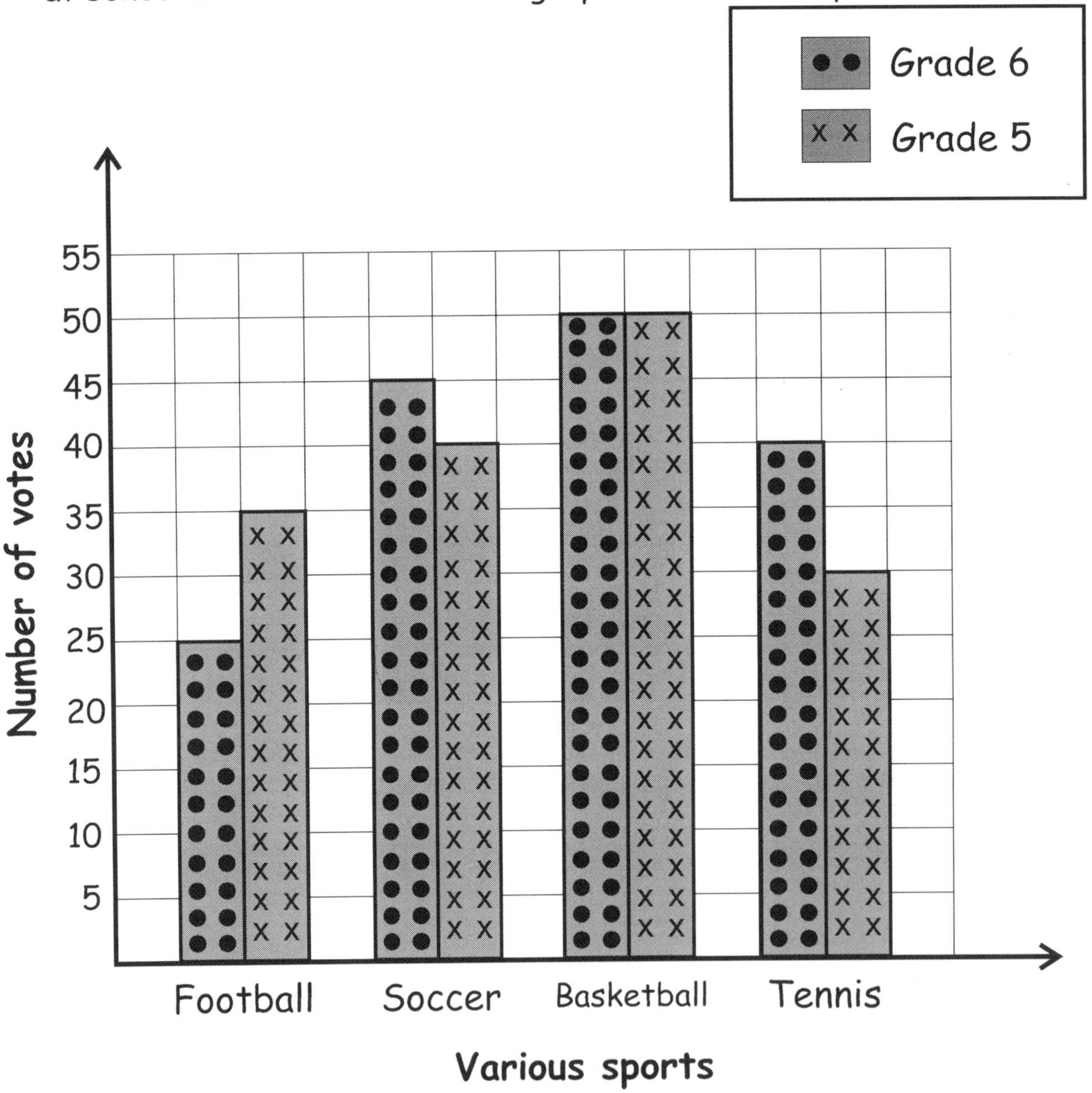

6. How many 5th grade students voted ?

(A) 140

(B) 50

(C) 165

(D) 135

7. How many 6th grade students voted ?

(A) 150 (B) 90

(C) 180 (D) 160

8. How many 6th grade students voted than 5th grade students ?

(A) 14 (B) 15

(C) 5 (D) 31

9. How many 6th grade students voted for soccer than 5th grade students ?

(A) 5 (B) 11

(C) 19 (D) 30

10. How many students voted for football all together ?

(A) 26 (B) 20

(C) 60 (D) 58

The population growth in Virginia across various years is plotted as below. Answer the questions 11-15

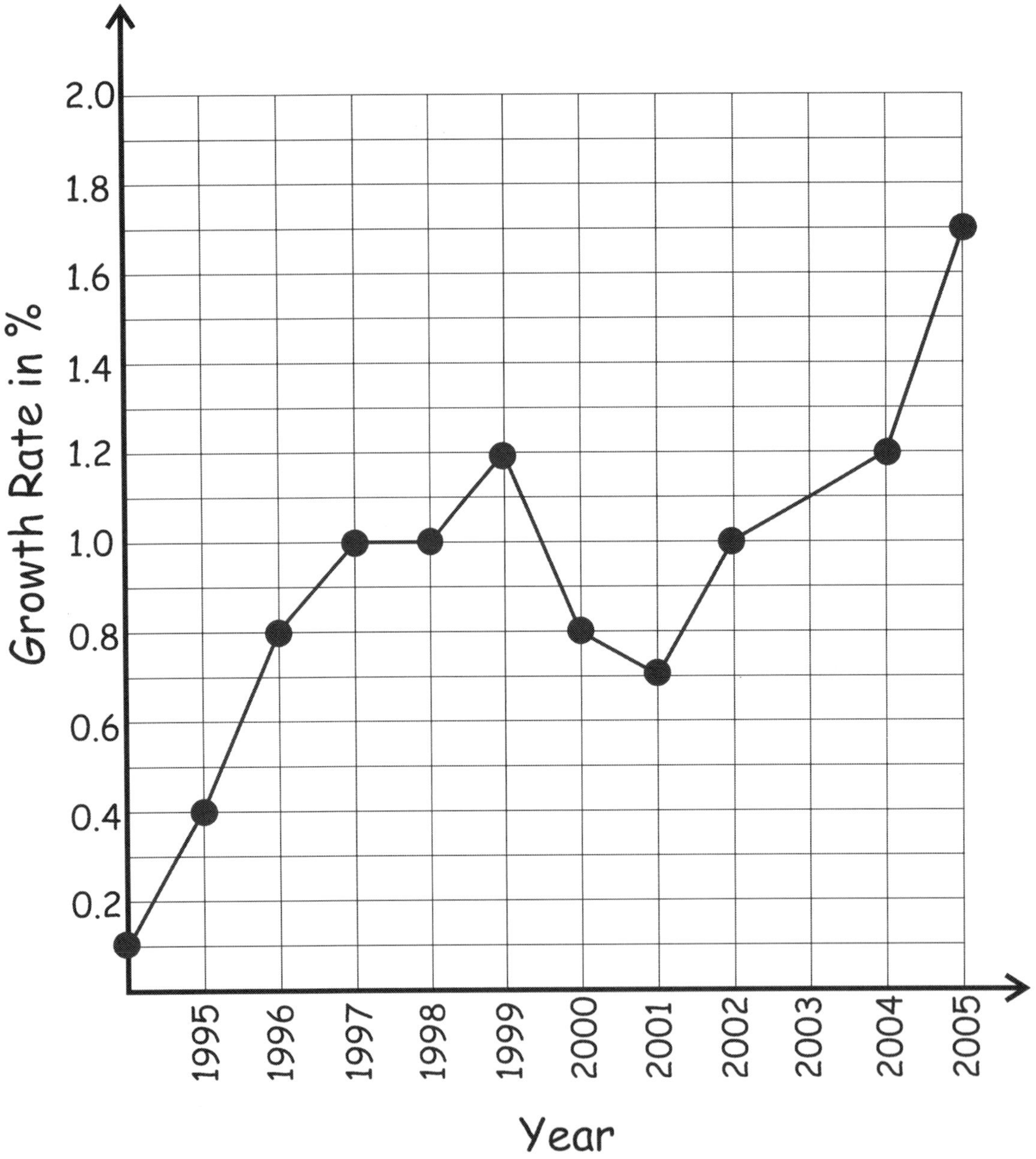

11. What is the population growth between 1995 and 1998 ?

(A) 1.0

(B) 0.6

(C) 0.8

(D) 1.4

12. What is the population growth between 2002 and 2004 ?

(A) 0.2 (B) 2.2

(C) 1.2 (D) 0.8

13. What is the decrease in growth between 1999 and 2001 ?

(A) 0.3 (B) 0.75

(C) 0.4 (D) 0.5

14. In which year the population growth is largest ?

(A) 2005 (B) 2002

(C) 1999 (D) 1995

15. What is the population growth between 1997 and 1998 ?

(A) 2 (B) 1

(C) 0.8 (D) 0

SECTION D

Andrew is analyzing sales of skate boards in his sports shop this year. Based in the below data, answer the questions 1-4.

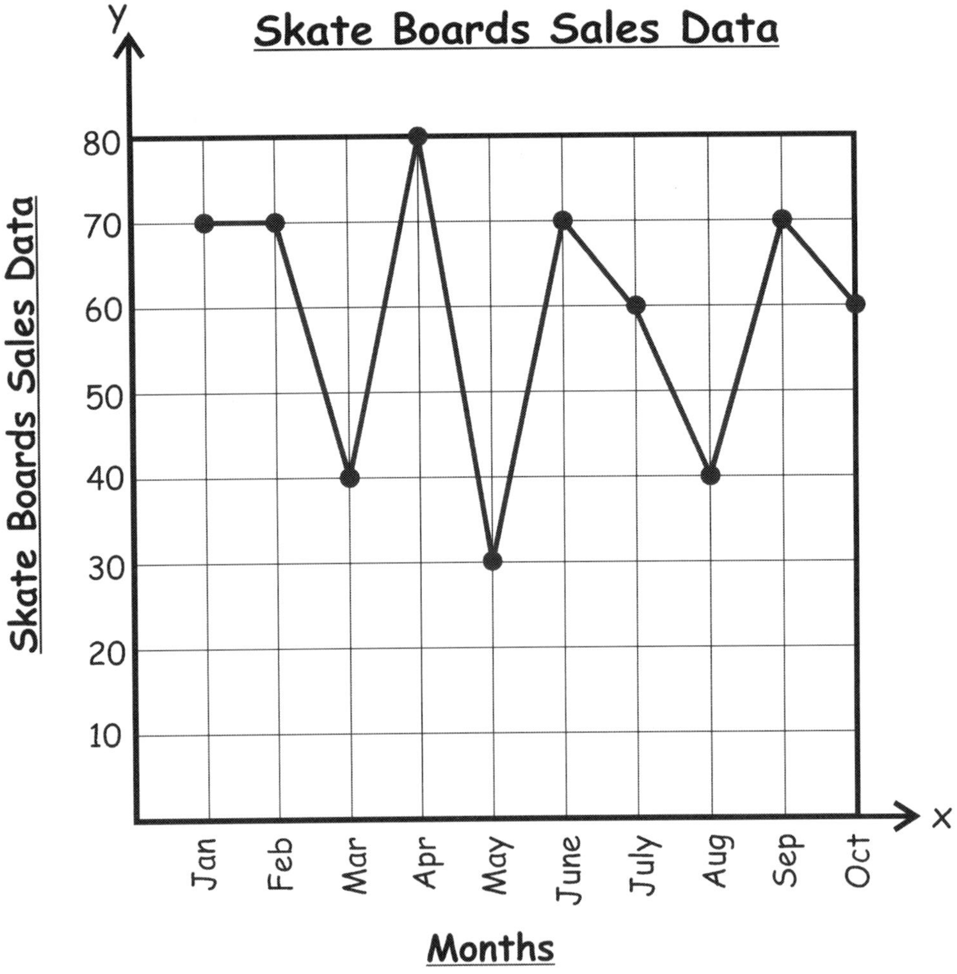

1. What is the median of the above data ?

(A) Median = 60 (B) Median = 65

(C) Median = 85 (D) Median = 70

2. What is the Mean of the above data ?

(A) Mean = 29 (B) Mean = 55

(C) Mean = 10.2 (D) Mean = 59

3. What is the Range of the above data ?

(A) 110 (B) 40

(C) 50 (D) 70

4. The area of a square whose side measures four units is added to the difference of eleven and nine divided by two. What is the total value ?

(A) 17 (B) 16

(C) 5 (D) 60

5. If twice the quantity x + 6 is divided by negative four, the result is 5. Find the number ?

(A) -18 (B) -16

(C) -13 (D) -0.5

6. Train A left the station traveling south one hour before a train B. Train B traveled in the opposite direction going 20 mph slower than Train A for 11 hours after which time the trains were 1390 mi. apart. Find the Train A's speed.

(A) 90 mph (B) 68 mph

(C) 75 mph (D) 70 mph

Analyze the graph given below. Answer questions 7-11

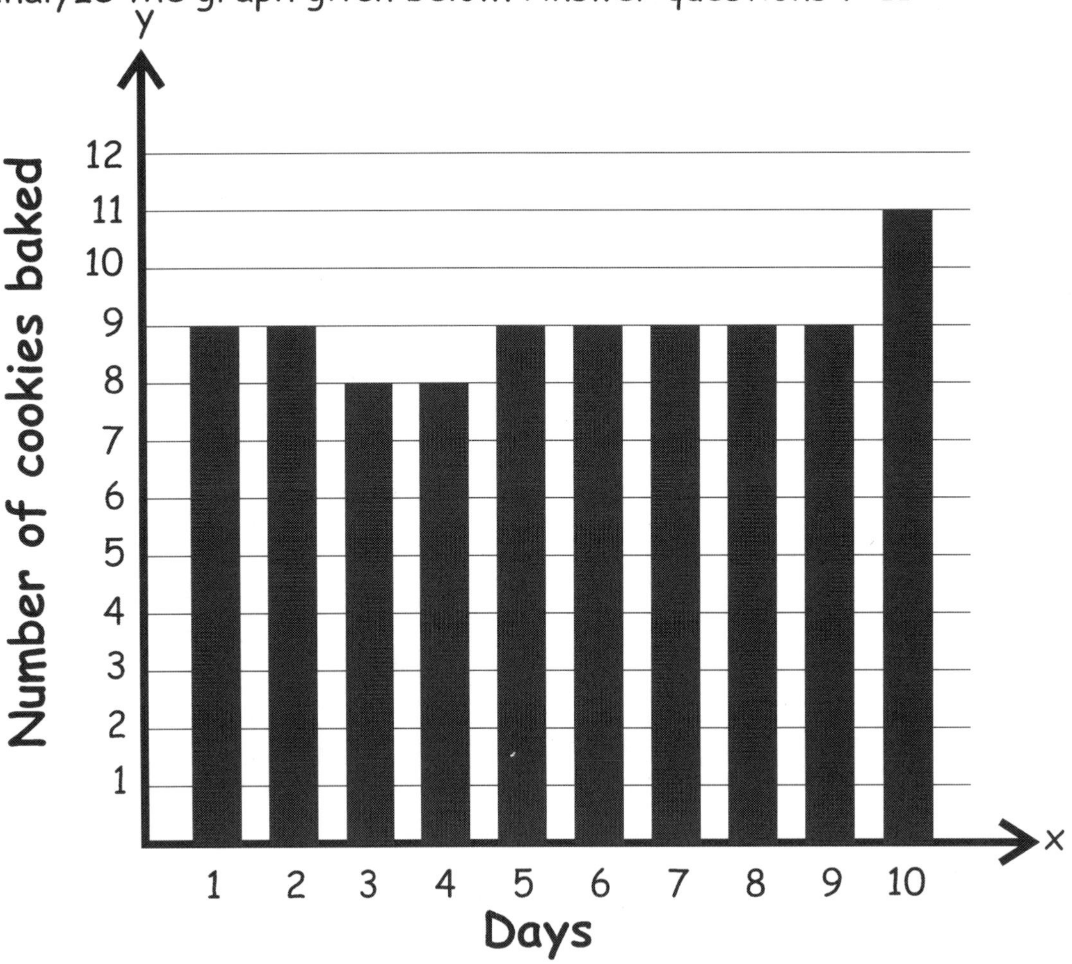

7. What is the median of the above data ?

(A) Median = 9 (B) Median = 8

(C) Median = 6 (D) Median = 11

8. What is the Mean of the above data ?

(A) Mean = 7 (B) Mean = 6

(C) Mean = 8.5 (D) Mean = 9.0

9. What is the Range of the above data ?

(A) 5 (B) 3

(C) 4 (D) 9

10. What is the Mode of the above data ?

(A) 2 (B) 5

(C) 9 (D) 8

11. Find $\angle WEF$ if $\angle DEF = 180^0$ and $\angle DEW = 23^0$

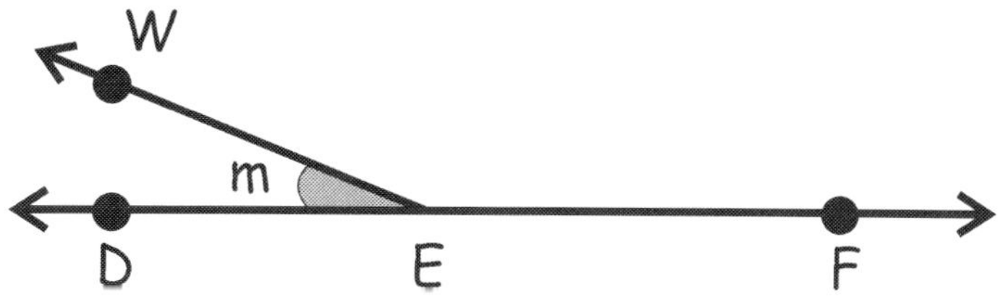

(A) 133^0 (B) 156^0

(C) 107^0 (D) 167^0

12. Evaluate $\frac{|2 - 8m|}{5} = 2$

(A) $\{-1, \frac{-4}{3}\}$

(B) $\{\frac{2}{11}, -3\}$

(C) $\{-1, \frac{3}{2}\}$

(D) $\{-1, \frac{-3}{2}\}$

13. Find the value of $\angle x$ from the figure below

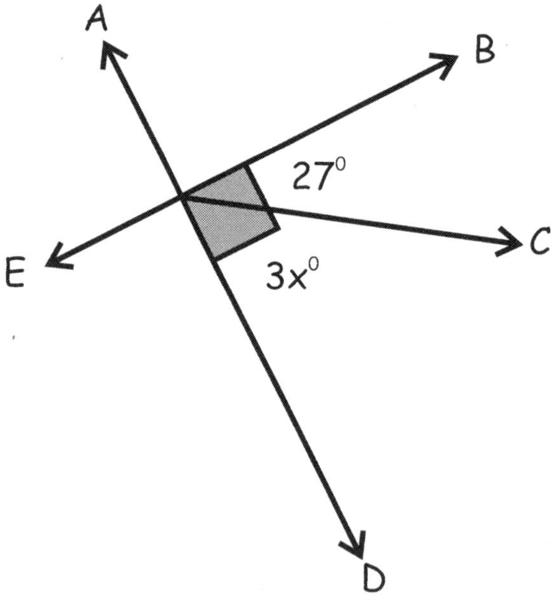

(A) 20°

(B) 21°

(C) 17°

(D) 25°

14. Evaluate x from the figure below

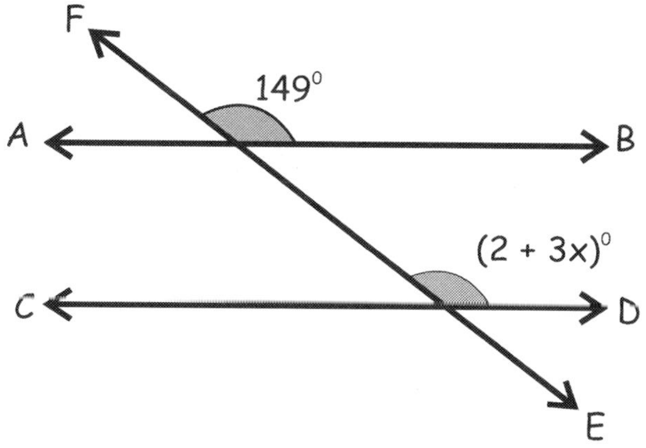

(A) 49°
(B) 52°
(C) 45°
(D) 61°

15. Find the slope of the straight line from the graph given below.

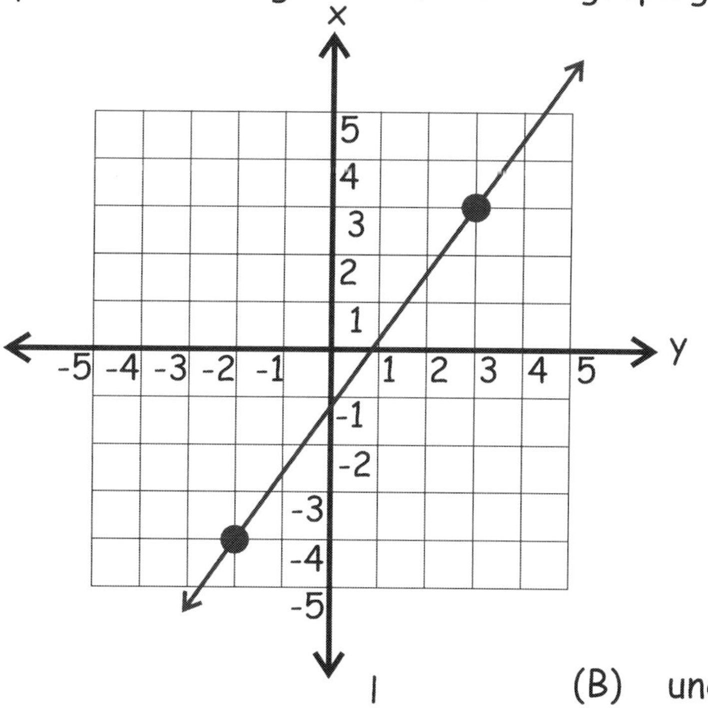

(A) 0
(B) undefined
(C) $\frac{-7}{5}$
(D) $\frac{7}{5}$

IAAT
PRACTICE TEST
TEST - 14

SECTION A

1. Find the percentage change from 61 to 100 rounding to the nearest tenth.

   (A) 63.9% decrease
   (B) 39% increase
   (C) 63.9% increase
   (D) 74.7% decrease

2. Original price of a dress is $ 50.00. Store offers a 10% discount. What is the sale price of the dress ?

   (A) $ 50.00
   (B) $ 5.00
   (C) $ 45.00
   (D) $ 40.00

3. Find the value of H satisfying the below equation

   $$-3 < H < 8$$

   (A) 0
   (B) 11
   (C) -4
   (D) 8

4. Find the value of H satisfying the below equation

   $$-13 < H < -7$$

   (A) -14
   (B) -40
   (C) 10
   (D) -5

5. Find the value of H satisfying the below equation

$$H*H - T = 35 \quad \text{where } T = 1$$

(A) 2
(B) 0
(C) 6
(D) 9

6. Find the value of H satisfying the below equation

$$2H + 7T = 11 \quad \text{where } T = 1$$

(A) 2
(B) 4
(C) 1
(D) 8

7. Simplify the below :

$$(3p - 8p^3) + (6p^3 - 5p)$$

(A) $-p^3 - p$
(B) $2p^3 + 2p$
(C) $-5p^3 + 2p$
(D) $-2p^3 - 2p$

8. Evaluate : $-8(7x-3)-8x=344$

(A) 11
(B) -5
(C) 0
(D) 9

9. Evaluate : $11 \div 12.5$

(A) 0.88

(B) 0.088

(C) 0.54

(D) 1.37

10. Evaluate : $\frac{x}{4} + 10 = 14$

(A) -16

(B) 12

(C) 8

(D) 16

11. Find the value of y given the points on straight line

D(-2,-2) and E(0,y) ; slope = $\frac{9}{2}$

(A) 10

(B) 19

(C) 7

(D) 27

12. Evaluate the below, where a = 6 and b = 7

$$a \times \frac{b - 3 + a}{2}$$

(A) 30

(B) 78

(C) 25

(D) 56

13. Simplify : $3\frac{2}{5} - \frac{1}{2}$

(A) $4\frac{1}{10}$

(B) $4\frac{31}{40}$

(C) $3\frac{17}{30}$

(C) $2\frac{9}{10}$

14. Write an algebraic expression for the below

A number cubed is less than 101

(A) $3^n \leq 101$

(B) $n^3 > 101$

(C) $3n = 101$

(D) $n^3 < 101$

15. Express the below verbally

$$\frac{n}{2} \geq 48$$

(A) half of n is greater than or equal to 48

(B) the difference of 2 and n is greater than or equal to 48

(C) half of 2 is greater than or equal to 48

(D) the second power of n is greater than or equal to 48

SECTION B

1. Which of the following expressions is equivalent to

$$a^6$$

(A) a x a x a x a

(B) a x a x a x a x a x a

(C) a x a x a x a x a

(D) a x a x a x a

Based on the data given below, answer the question 2-4

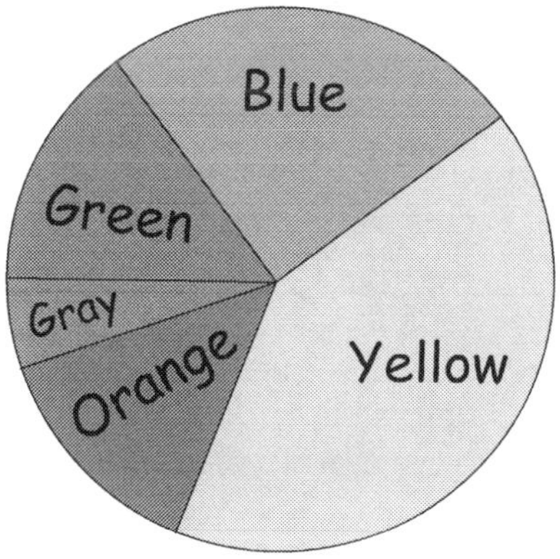

2. Which color is most voted ?

(A) Yellow

(B) Green

(C) Orange

(D) Black

3. Which color is least voted ?

(A) Orange

(B) Blue

(C) Green

(D) Gray

4. Which two colors have equal votes?

(A) Orange and Green (B) Gray and Orange

(C) Blue and Yellow (D) Yellow and Green

5. Which of the below point lies in second quadrant ?

(A) (-2,3) (B) (2,-3)

(C) (2,3) (D) (-2,-3)

6. Which of the below point lies in first quadrant ?

(A) (-5,-5) (B) (5,0)

(C) (0,-5) (D) (5,5)

7. Simplify the below

$$(p \times p \times p)(q \times q \times q \times q)$$

(A) $p^3 x q^4$ (B) $p^2 x q^4$

(C) $p^4 x q^3$ (D) $p^6 x q^7$

8. The ratio of pencils to erasers in a box is 2:3. If there are 10 pencils then, how many erasers are there ?

(A) 15 (B) 11

(C) 23 (D) 30

9. Helen has 10 black shirts for every 5 blue shirts in her closet. She now has 8 black shirts. Which of the below proportion can be used to find the number of blue shirts she has now ?

(A) $\frac{10}{5} = \frac{x}{8}$ (B) $\frac{10}{5} = \frac{8}{x}$

(C) $\frac{5}{10} = \frac{8}{x}$ (D) $5x = 80$

10. Which of the following has the highest value ?

(A) $|-21|$ (B) $|-36|$

(C) $-|1525|$ (D) $|-93|$

11. Evaluate for p

$$5 \times 32 = (5 \times 30) + (5 \times p)$$

(A) 2 (B) 5

(C) 0 (D) 1

12. A particular house is 17.6 m wide. A model of it was built with a scale of 2.5 cm : 11 m. How wide is the model ?

(A) 77.4 cm (B) 4 cm

(C) 6.8 cm (D) 2.1 cm

13. The below equation is an example of

$$(2a + 5b) + 7c = 2a + (5a + 7c)$$

(A) Distributive property (B) Commutative property

(C) Associative property (D) Symmetric property

14. How many inches are in three feet ?

(A) 18
(B) 30
(C) 32
(D) 36

15. Which of the below equations is represented by the figure give below ?

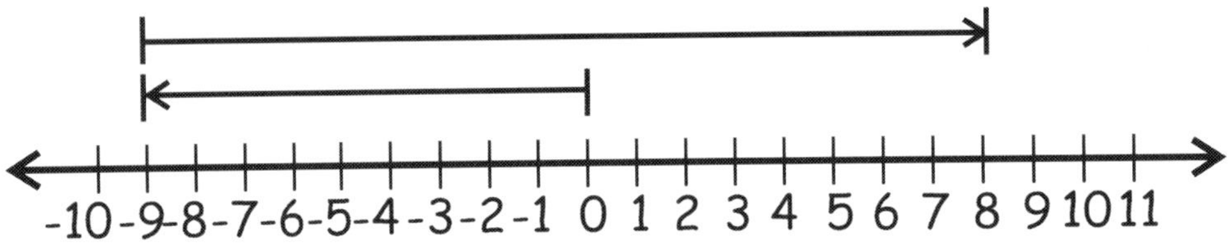

(A) -17 + 0
(B) -17 + 9
(C) -9 + 17
(D) -9 + 9

SECTION C

Bull Run elementary school students voted on various Ice cream flavors, at school as shown in the below graph. Answer the questions 3-7

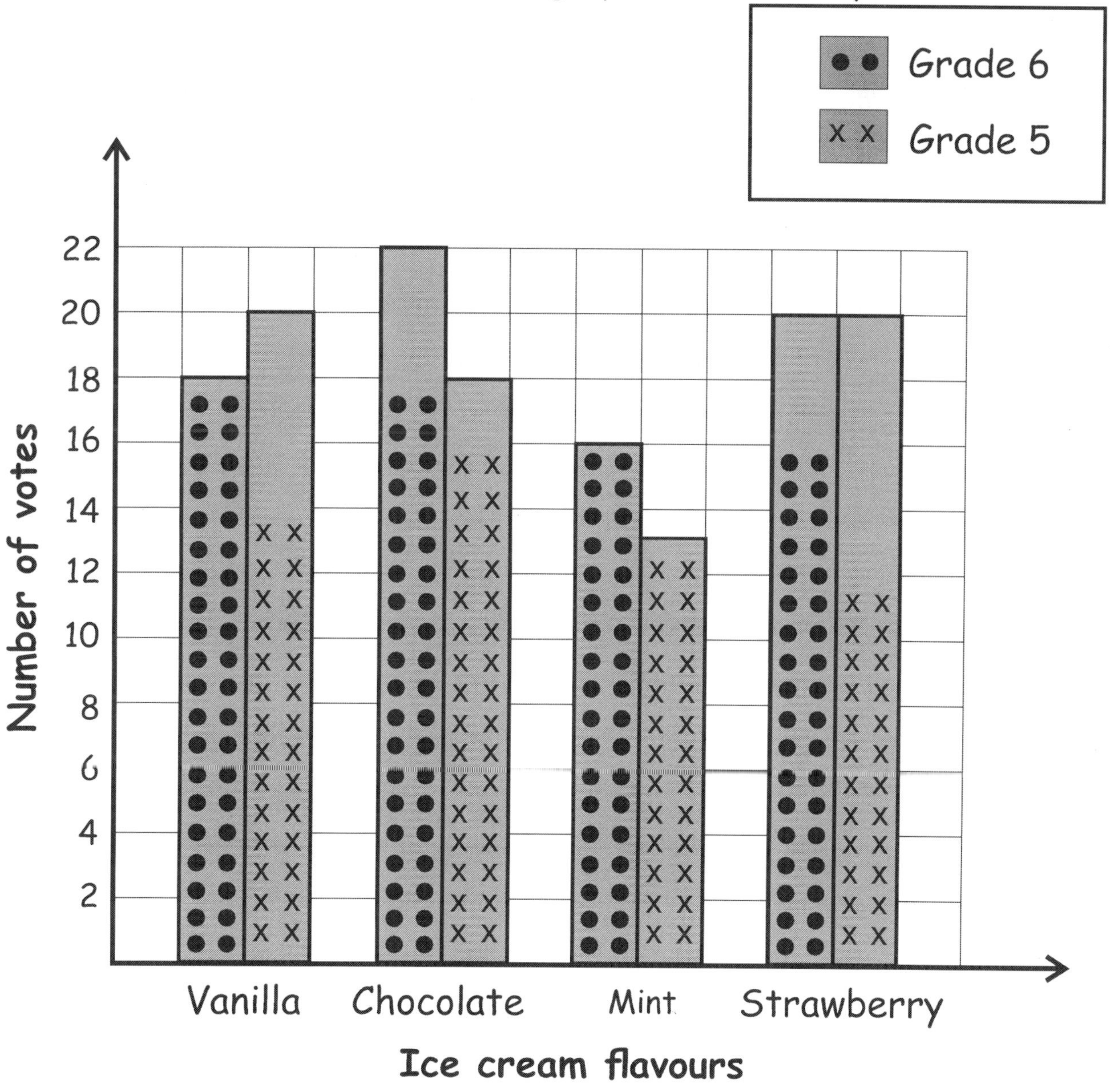

1. How many 5th grade students voted ?

(A) 71

(B) 81

(C) 77

(D) 95

2. How many 6th grade students voted ?

(A) 61 (B) 65

(C) 70 (D) 76

3. How many 6th grade students voted than 5th grade students ?

(A) 4 (B) 5

(C) 11 (D) 29

4. How many 6th grade students voted for Chocolate flavor than 5th grade students ?

(A) 4 (B) 17

(C) 5 (D) 0

5. How many students voted for Mint flavor all together ?

(A) 26 (B) 20

(C) 29 (D) 38

Jack bikes on every weekend, and the data is plotted as below.
Answer the questions 8-12

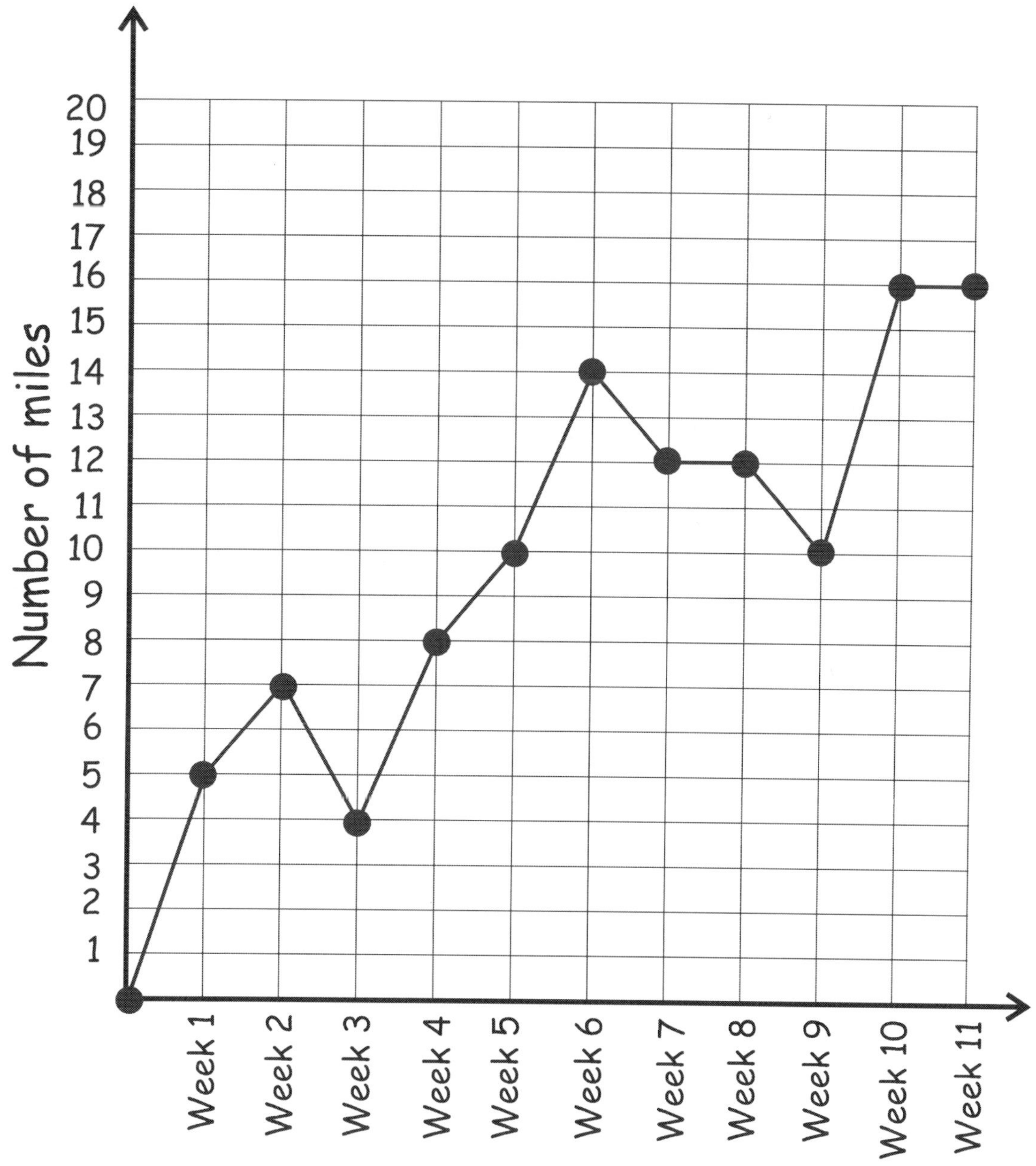

6. How many miles did he ran between Week 9 and week 10 ?

(A) 6

(B) 5

(C) 8

(D) 21

7. How many miles did he ran between Week 3 and Week 6 ?

(A) 31 (B) 12

(C) 19 (D) 10

8. How many miles did he ran between Week 4 and Week 11 ?

(A) 3 (B) 6

(C) 8 (D) 15

9. In which week he covered the largest number of miles ?

(A) Weeks 10 and 11 (B) Weeks 3 and 6

(C) Weeks 7 and 8 (D) Weeks 6 and 9

10. How many miles did he ran between week 1 and week 4 ?

(A) 7 (B) 3

(C) 13 (D) 1

Students at creative art center voted for their most favorite color as shown in the below graph. 126 students like blue color. Based on this data answer the questions 13-15

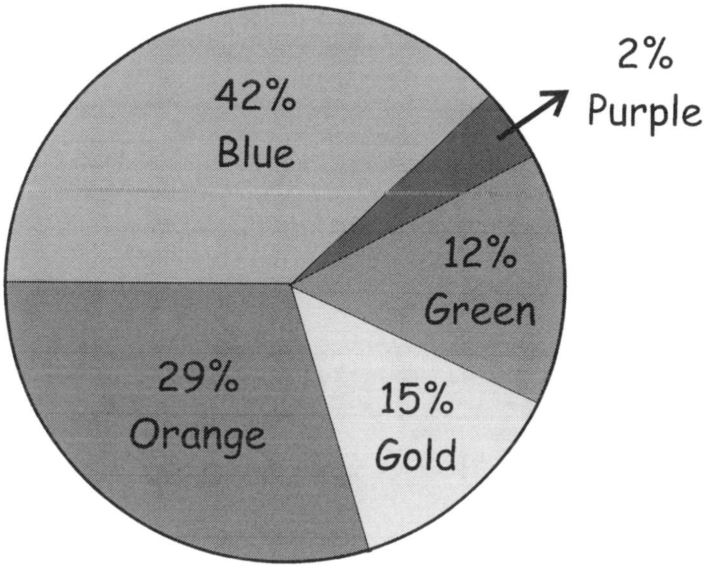

11. How many students are there in creative art center ?

(A) 150 (B) 175

(C) 225 (D) 300

12. How many students voted for Gold color ?

(A) 45 (B) 23

(C) 10 (D) 90

13. How many students Voted for Orange color ?

(A) 55 (B) 63

(C) 87 (D) 79

14. John has 30 pages to read for history class tonight. He decided that he would take a break when he finished reading 70% of the pages assigned. How many pages must he read before he takes a break ?

(A) 7 (B) 11

(C) 21 (D) 18

15. The sum of two consecutive even integers is 126. What are the integers ?

(A) 62,64 (B) 62.63

(C) 60,66 (D) 60,64

SECTION D

1. Store A sells dresses of various sizes as given below. Which of the below graphs shows the same data ?

Dress sizes

5 , 5 , 6 , 6 , 6.5 , 7.5 , 8 , 8 , 8 , 8.5 , 9 , 11 , 12.5 , 13 , 13

(A)

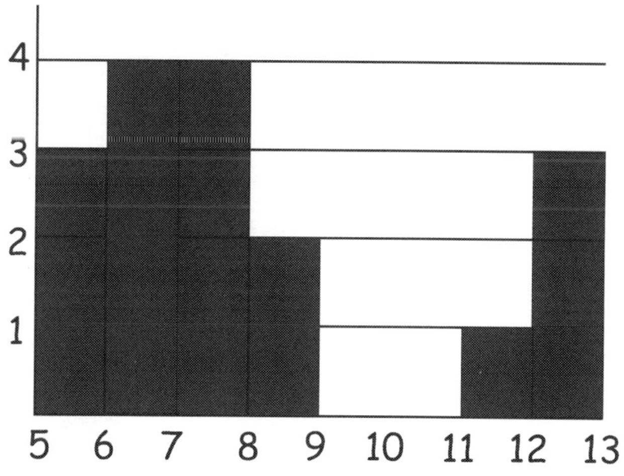

(B)

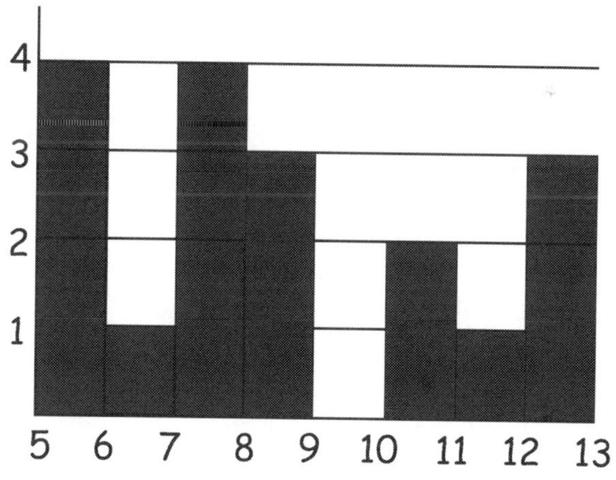

(C)

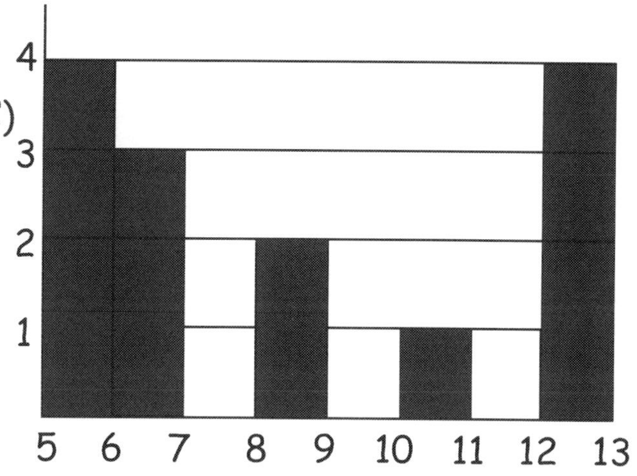

(D)

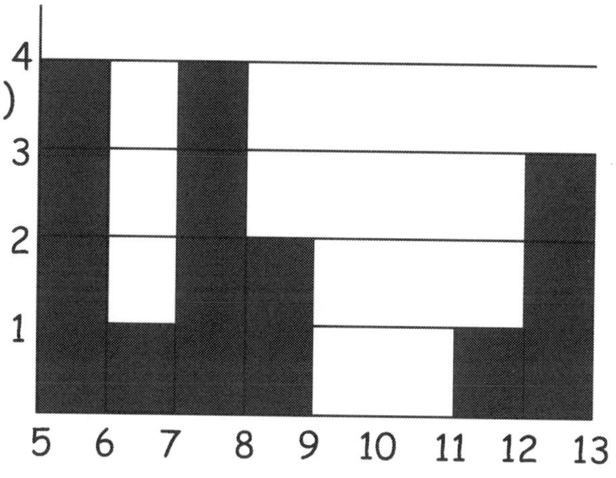

2. Which graph is plotted as per the sales data given below ?

Plants	# Sold
Daisies	40
Roses	45
Marigold	50
Dahlia	65
Lilac	80

Plants	# Sold
Petunia	80
Peony	80
Jasmine	85
Orchids	100
Deffodil	100

(A)
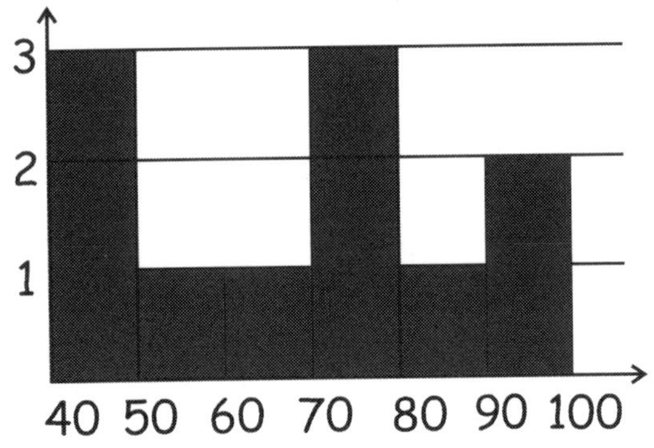

(B)
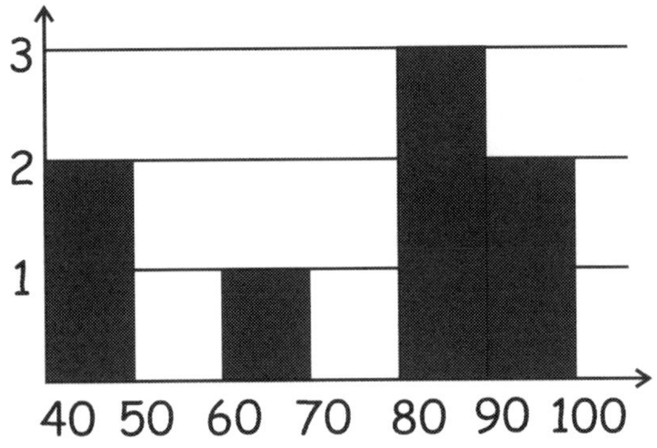

(C)
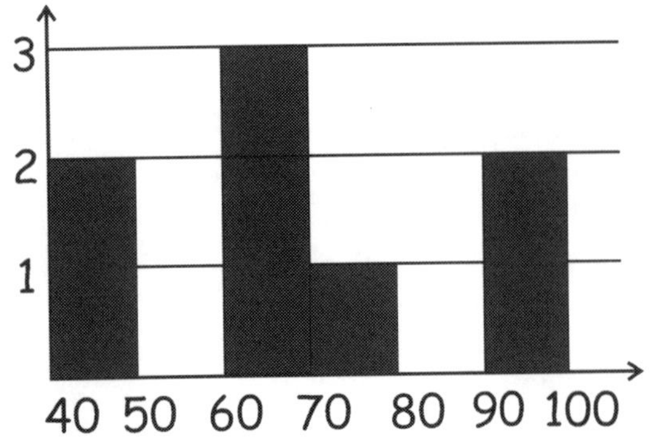

(D)
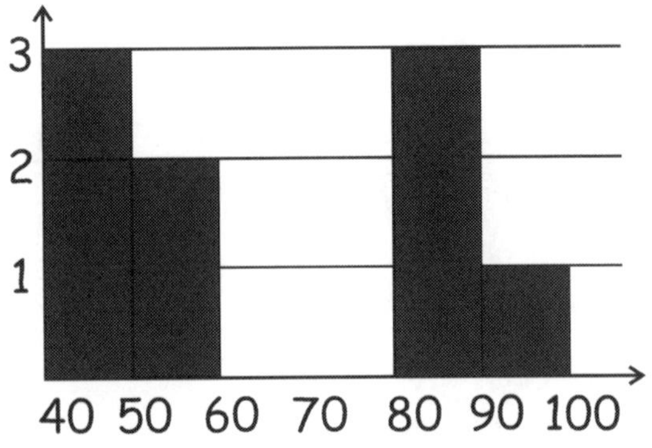

3. Which graph is plotted as per the data given below ?

Names	Scores
Sophia	74
Elena	74
Sarah	75
Nova	80
Camila	81

Names	Scores
Avery	81
Harper	81
Emma	82
Ruby	82
Hannah	84

(A)
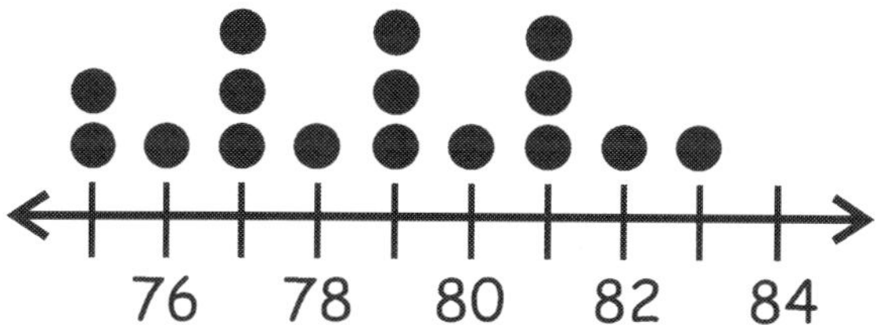

(B)
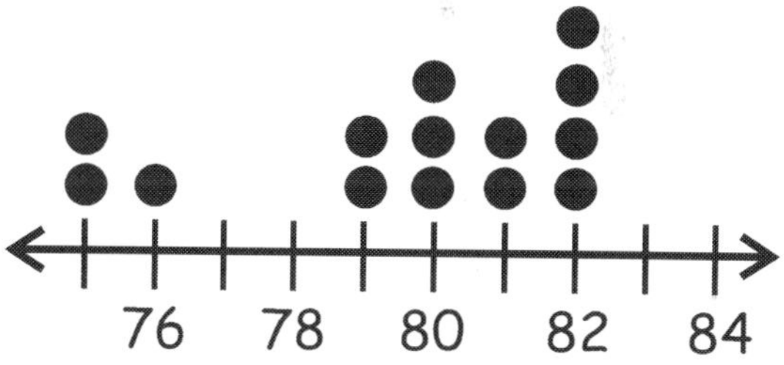

(C)
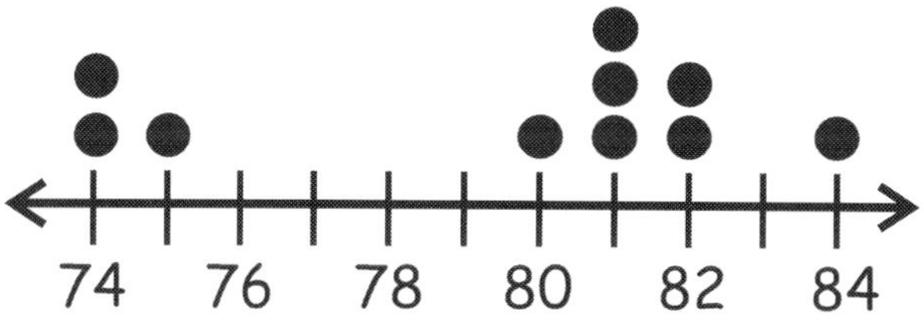

(D)
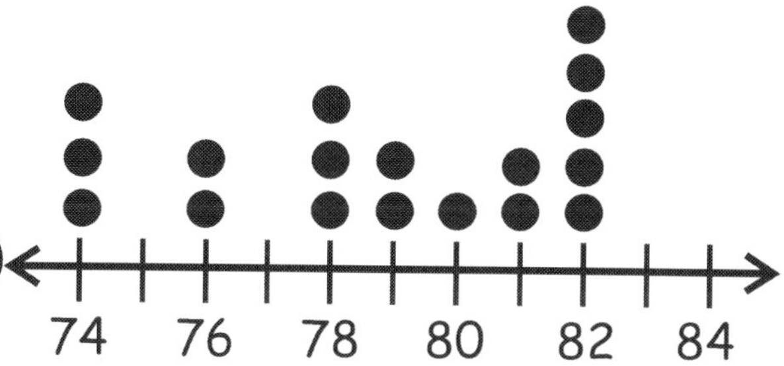

4. Which graph is plotted as per the data given below ?

State	Years
Florida	50
California	53
Atlanta	59
Virginia	69
Mary land	71
Washington D.C	74

State	Years
Boston	76
North Carolina	77
Georgia	81
Cansas	81
New Jersey	83

(A)

Stem	Years
5	1 6
6	6 8
7	1 4 4 7
8	1 1 5

(B)

Stem	Years
5	0 3 9
6	9
7	1 4 6 7
8	1 1 3

(C)

Stem	Years
5	0 7 8
6	
7	2 3 3 6 7 7 9
8	3

(D)

Stem	Years
5	5 6 7
6	8
7	9 9
8	0 2 4 5 5

5. Based on the given data plot a Box and Whisker plot

Names	Score
Skylar	2
Madeline	2
Claire	4
Violet	8
Hazel	9
Stella	10
Piper	12
Lydia	18
Quin	26

(A)

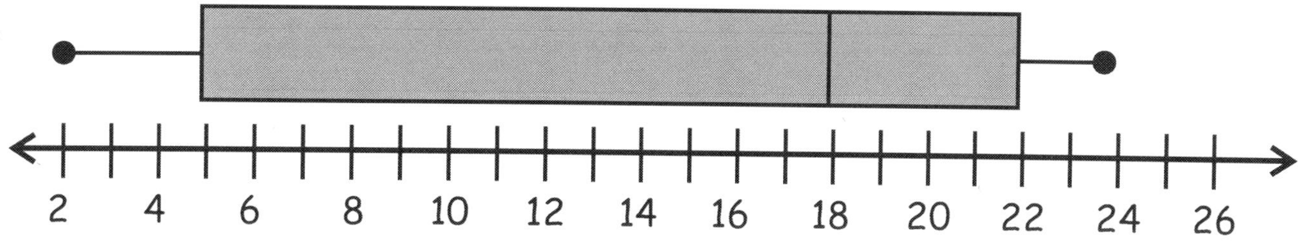

(B)

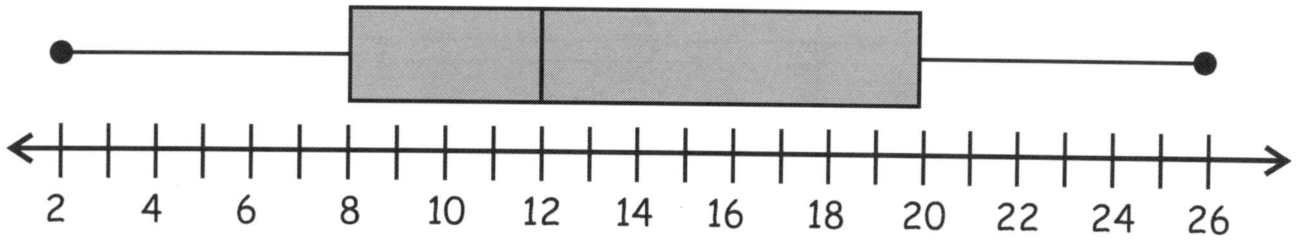

(C)

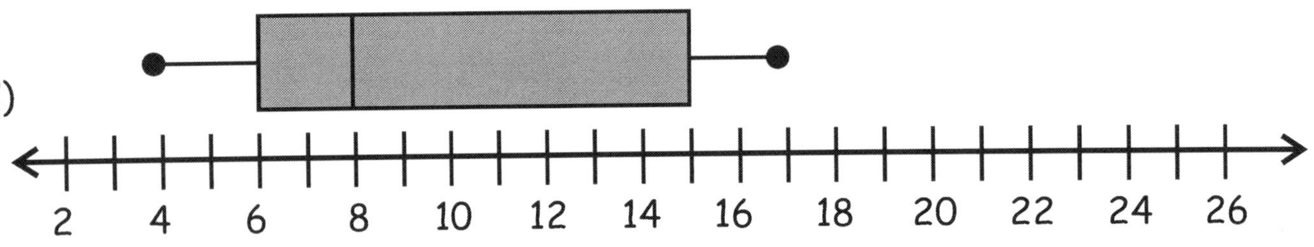

(D)

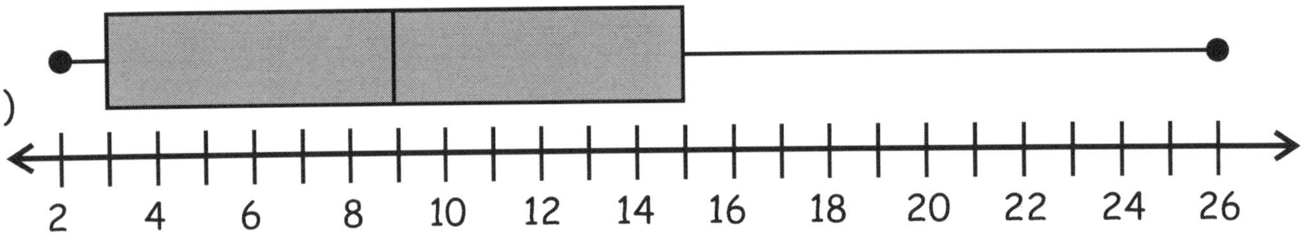

6. Given y = 5x + 7 when x increases what happens to y ?

(A) Increases

(B) Decreases

(C) No change

(C) None

7. Find the additive inverse of 75

(A) 1

(B) -74

(C) -75

(D) 0

8. Find the irrational number from the given options given below ?

(A) $\sqrt{4}$

(B) $\sqrt{100}$

(C) $\sqrt{124}$

(D) $\sqrt{81}$

Based on the graph given below, Answers the questions 1 - 4

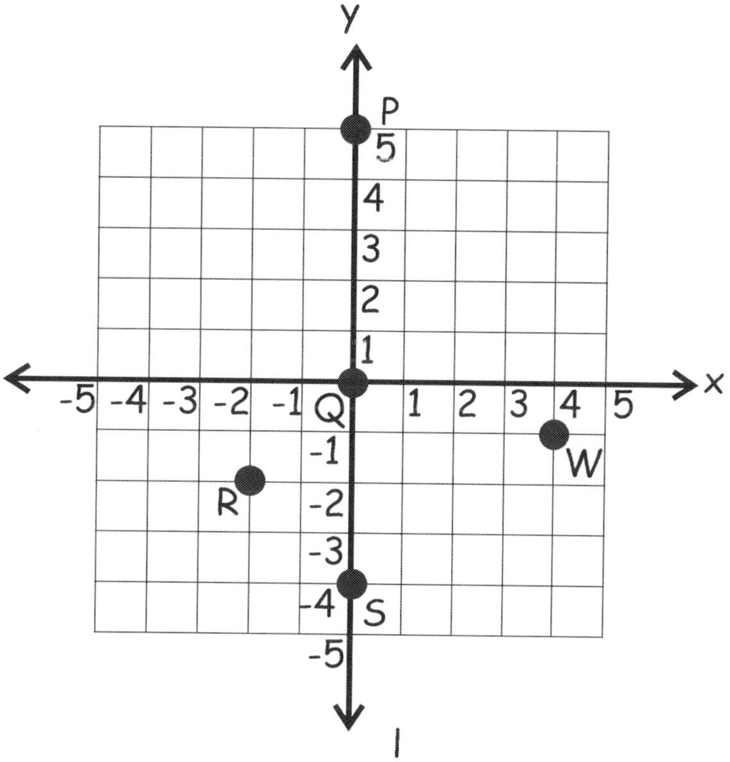

9. a. The point R lies in which quadrant ?

(A) Quadrant I
(B) Quadrant II
(C) Quadrant III
(D) Quadrant IV

b. The point W lies in which quadrant ?

(A) Quadrant I
(B) Quadrant II
(C) Quadrant III
(D) Quadrant IV

10. Which of the below represents the coordinates of point P ?

(A) (0,5) (B) (0,0)

(C) (5,0) (D) (0,-5)

11. Which of the below represents the coordinates of Q ?

(A) (2,4) (B) (0,2)

(C) (0,0) (D) (3,1)

12. $|8 - b| + 8 = 24$

(A) {-8,24} (B) $\frac{8}{9}$

(C) $\frac{2}{5}$ (D) $\{\frac{8}{9}, -2\}$

13. $-9|6b - 2| = -36$

(A) {-8,-4} (B) 1

(C) {12,-8} (D) $\{1, -\frac{1}{3}\}$

14. Rotate the $\triangle PQR$, 90° Counter Clockwise. Find the coordinates of each vertex after the transformation.

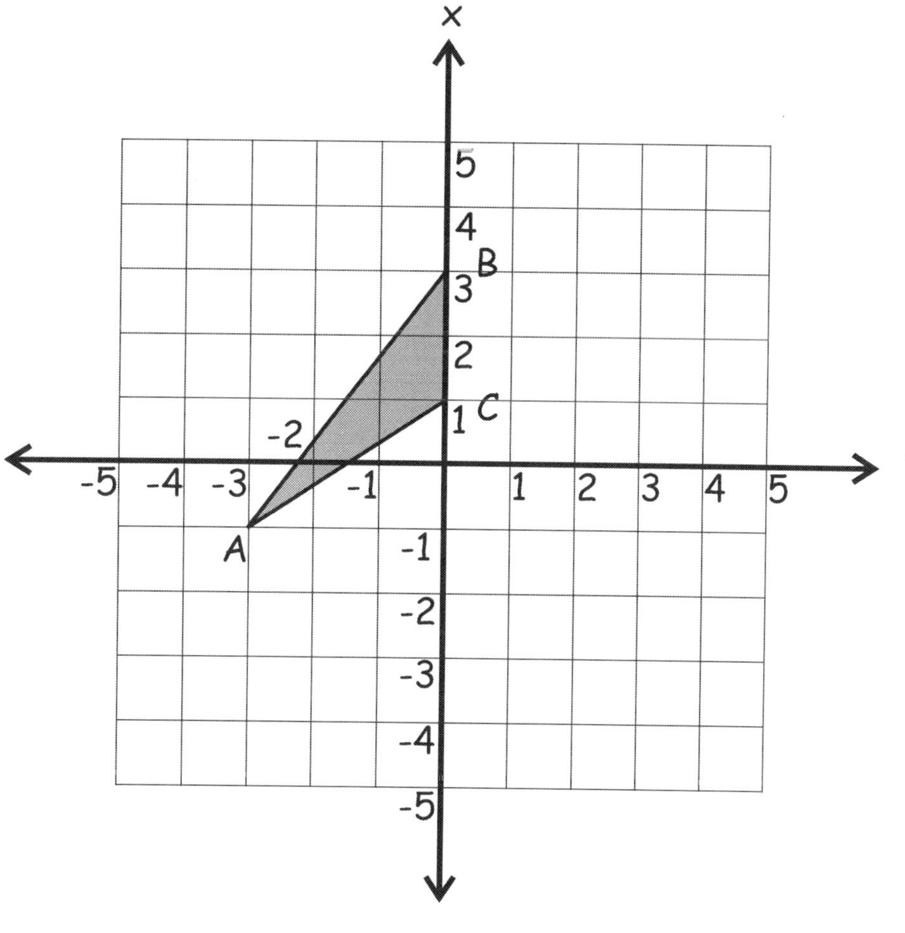

(A) A′(5,1),B′(1,5),C′(-2,3)

(B) A′(1,3),B′(-3,0),C′(1,0)

(C) A′(1,-3),B′(-3,0),C′(-1,0)

(D) A′(1,-3),B′(0,-3),C′(0,-1)

15. Rotate the $\triangle PQR$, 90° Counter Clockwise. Find the coordinates of each vertex after the transformation.

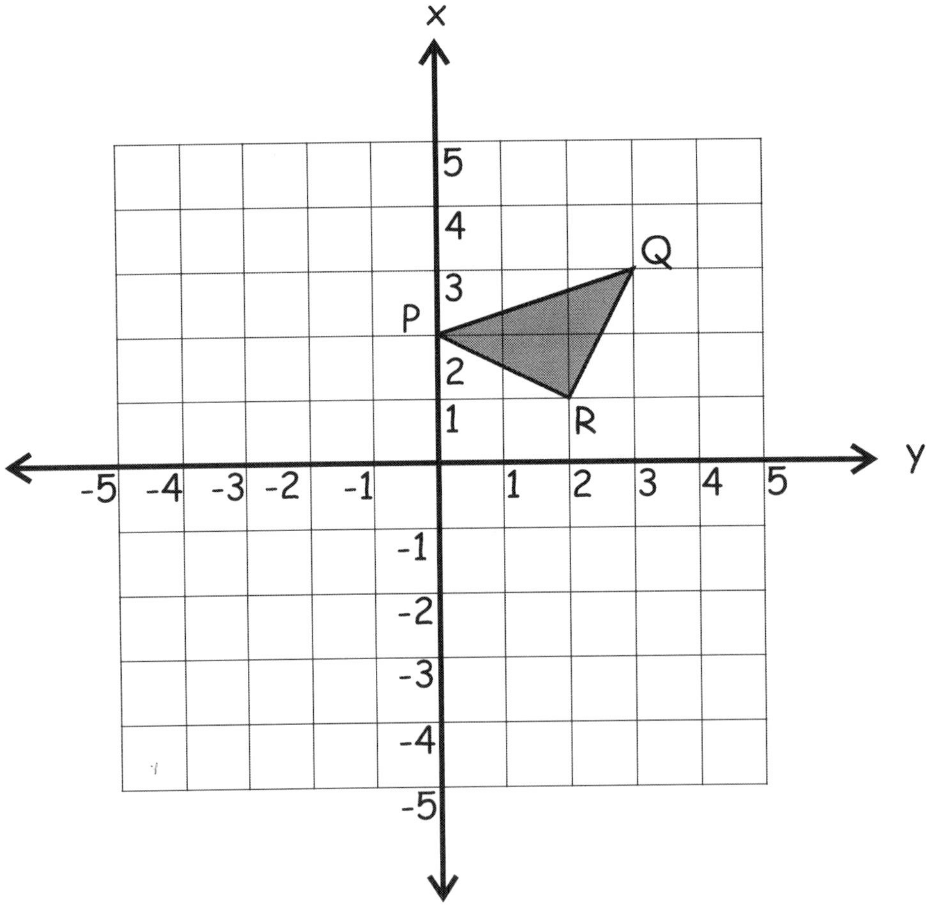

(A) P'(-4,1),Q'(1,2),R'(2,0)

(B) P'(2,0),Q'(3,3),R'(1,-2)

(C) P'(0,-2),Q'(-3,-3),R'(-2,-1)

(D) P'(2,0),Q'(3,-3),R'(1,-2)

IAAT
PRACTICE TEST
Answer Keys

Test 7 Answer Keys

Question No	Section A	Section B	Section C	Section D
1.	A	C	A	A
2.	D	C	B	C
3.	B	B	D	B
4.	A	C	B	B
5.	B	B	A	A
6.	B	D	B	D
7.	D	C	B	(a) A (b) B
8.	D	D	A	(a) B (b) A
9.	B	B	D	A
10.	D	C	C	D
11.	A	A	C	(a) C (b) D
12.	A	C	D	(a) B (b) C
13.	C	C	D	A
14.	A	D	D	B
15.	A	B	C	A

Test 8 Answer Keys

Question No	Section A	Section B	Section C	Section D
1.	D	D	A	A
2.	C	A	C	D
3.	A	C	D	C
4.	C	D	C	D
5.	B	D	B	A
6.	D	B	A	C
7.	A	C	A	(a) A (b) D
8.	D	B	D	(a) B (b) D
9.	A	C	D	C
10.	A	D	C	A
11.	D	B	C	(a) A (b) B
12.	B	B	B	(a) B (b) C
13.	C	A	B	D
14.	B	C	B	B
15.	B	D	B	A

Test 9 Answer Keys

Question No	Section A	Section B	Section C	Section D
1.	B	B	B	A
2.	A	A	D	C
3.	D	B	A	D
4.	B	D	B	A
5	A	B	D	B
6.	B	A	C	B
7.	B	C	B	(a) D (b) A
8.	D	A	A	(a) B (b) A
9.	A	C	C	C
10.	A	A	B	D
11.	C	D	D	(a) B (b) D
12.	B	B	D	(a) A (b) A
13.	B	C	A	D
14.	D	B	B	D
15.	A	D	B	C

Test 10 Answer Keys

Question No	Section A	Section B	Section C	Section D
1.	C	D	A	D
2.	A	B	B	A
3.	D	D	D	C
4.	D	A	B	D
5.	B	D	A	D
6.	D	C	A	D
7.	B	A	D	(a) B (b) A
8.	B	C	B	(a) C (b) D
9.	A	B	B	A
10.	A	C	A	D
11.	C	A	D	(a) C (b) C
12.	D	A	B	(a) A (b) D
13.	C	C	A	A
14.	B	C	C	C
15.	D	A	A	C

Test 11 Answer Keys

Question No	Section A	Section B	Section C	Section D
1.	A	D	A	(a) C (b) D
2.	A	A	C	A
3.	C	A	A	A
4.	B	D	B	B
5.	C	C	B	D
6.	D	A	C	D
7.	D	B	A	D
8.	D	D	D	D
9.	B	A	D	C
10.	D	D	A	A
11.	B	A	D	D
12.	B	A	B	A
13.	B	B	A	C
14.	C	D	C	B
15.	C	C	B	D

Test 12 Answer Keys

Question No	Section A	Section B	Section C	Section D
1.	D	A	D	(a) B (b) A
2.	C	D	B	D
3.	A	B	A	C
4.	C	B	C	B
5.	C	A	D	B
6.	C	D	A	D
7.	C	B	B	C
8.	A	B	C	D
9.	A	B	D	D
10.	D	D	D	B
11.	C	D	A	C
12.	D	D	A	A
13.	C	B	D	C
14.	A	A	C	C
15.	D	C	A	A

Test 13 Answer Keys

Question No	Section A	Section B	Section C	Section D
1.	D	A	A	B
2.	C	C	D	D
3.	A	B	C	C
4.	A	A	D	A
5.	A	C	B	B
6.	B	B	C	D
7.	C	D	D	A
8.	A	A	C	D
9.	D	D	A	B
10.	A	A	C	C
11.	B	C	B	B
12.	D	B	A	C
13.	A	C	D	B
14.	B	B	A	A
15.	D	B	D	D

Test 14 Answer Keys

Question No	Section A	Section B	Section C	Section D
1.	C	C	A	D
2.	C	A	D	A
3.	A	D	B	C
4.	D	A	A	B
5.	C	A	C	D
6.	A	D	A	A
7.	D	A	D	C
8.	B	A	C	C
9.	A	B	A	(a) C (b) D
10.	D	D	B	A
11.	C	A	D	C
12.	A	B	A	A
13.	C	C	C	D
14.	D	D	C	C
15.	A	C	A	D

Made in United States
Orlando, FL
05 November 2026

90229838R00111